Masks
of
the
Universe

Masks
of
the
Universe

Edward Harrison

University of Massachusetts, Amherst

Macmillan Publishing Company
New York
Collier Macmillan Publishers
London

Copyright © 1985 by Macmillan Publishing Company
A Division of Macmillan, Inc.

Macmillan Publishing Company
866 Third Avenue, New York, NY 10022

Collier Macmillan Canada, Inc.

Printed in the United States of America

printing number
2 3 4 5 6 7 8 9 10

Library of Congress Cataloging in Publication Data
Harrison, Edward Robert.
  Masks of the universe.
  Bibliography: p.
  1. cosmology. I. Title.
QB981.H324 1985   523.1   84-23397
ISBN 0-02-948780-3

Contents

Preface

At first I thought this book would take me only a few months to write. After all, the basic idea was very simple, and only a few words should suffice to make it clear and convincing. But soon this illusion was shattered. A few months grew into 3 years, and now I realize that 30 years would not suffice. But enough! Other work presses, and life is too short.

This book, I regret to say, offers no comfort to earnest seekers of ultimate truth. It has no glib answers to the eternal questions, no glossy message on the nature of reality. Those who seek an easily purchased peace of mind in a world of dissonant tumult must inquire elsewhere. The first chapter, *Introducing the Masks*, explains what the book is about.

I take the view that we, and we alone, are the makers and shakers of the universes in which we live. The comfort I offer, if any, is of the kind that once sustained the medieval mystics who, while aware of much knowledge, had the sense to admit that they lived in "the cloud of unknowing." The message I offer, if any, is that the mysteries are vastly more profound and the mind vastly more potent than we commonly imagine, perhaps more than we can ever imagine.

No person can live as a human being, particularly as a member of a society, unless equipped with grand ideas concerning the world around. These grand ideas—or cosmic formulations—establish the universe in which that person lives. The universes that we create and in which we live determine the meaning and purpose of life. Wherever we find a society of intelligent beings (not necessarily intelligent by our standards), there we find a rational universe (but not necessarily rational by our standards); wherever we find a universe, there we find a society.

The universes, without which human beings could not live as intelligent beings, are the masks of the Universe. The unmasked Universe itself, however, remains forever unknown.

I hold the view that we alone create and organize the universes in which we live, and no universe is or ever can be the Universe. Yesterday a universe, today another, and tomorrow yet another. Each universe or mask presents a conceptual scheme that organizes human thoughts and shapes human understanding. Generally, within each universe the end to the search for all knowledge at last looms in sight. Each universe in its day flourishes as an awe-inspiring, self-consistent scheme of thought, yet each is doomed to be superseded by another and perhaps grander scheme.

I offer the thoughts in this book as nothing more than speculations. No doubt, even in this insecure age in constant peril, they will shock many persons and provoke strong disagreement. They dynamite all claims to absolute knowledge, explode all bodies of wisdom. Though readers may disagree, I shall not feel dismayed, provided they seek and gain their own views by sincere thought. According to my precepts, the thoughts expressed in this book are themselves ephemeral, factually superficial, and idealistically perishable.

I am grateful to the University of Massachusetts for a Faculty Fellowship, which enabled me to complete this work, and to the Institute of Astronomy, Cambridge University, for hospitality.

I am indebted to Vere Chappel, John Roberts, Carl Swanson, Oswald Tippo, and Peter Webster for their comments on certain ideas. To Michael Arbib, Thomas Arny, Jay Demerath, Laurence Marschall, Seymour Epstein, Gordon Sutton, David Van Blerkom, and Richard Ziemacki I express my gratitude for their comments regarding certain chapters. All, of course, remain entirely blameless. Most of all I appreciate the many thought-provoking remarks by Leroy Cook, who adopted an attitude of interested and yet amused detachment. Finally, I acknowledge the scathing and often devastating criticisms of Peter and June Harrison.

To my wife, Photeni, I dedicate *The Masks of the Universe*.

Masks
of
the
Universe

1

Introducing the Masks

The theme of this book is that the universe in which we live, or think we live, is mostly a world of our own making. The underlying idea rests on the distinction between *Universe* and *universes*. It is a simple idea having many consequences.

The Universe is everything. What it is in its own right, independent of our changing opinions, we never know. The Universe is all-inclusive and includes us; we are a part or an aspect of the Universe experiencing and thinking about itself.

What is the Universe? Logan Smith in *Trivia* expressed the nontrivial thought: "I awoke this morning ... into the daylight, the furniture of my bedroom—in fact into the well-known, often-discussed, but to my mind, as yet unexplained Universe." Were he alive today, he would still be waiting for an explanation. It is the oldest question, to which Socrates in his ironic manner replied, "I know nothing except the fact of my ignorance." I can think of no better answer.

The universes are our models of the Universe. They are great schemes of intricate thought—grand cosmic pictures—that rationalize human experience; these universes harmonize and invest with meaning the rising and setting Sun, the waxing and waning Moon, the jeweled lights of the night sky, the landscape of rocks and trees and clouds. Each universe is a self-consistent system of ideas, marvelously organized, interlacing most of what is perceived and known. *A universe is a mask fitted on the face of the unknown Universe.*

Wherever we find a human society, however primitive, there is a universe; and wherever we find a universe, of whatever kind, there is a society; both go together, and the one does not exist without the other. Each universe coordinates and unifies a society, enabling its members to communicate their thoughts and share their experiences. Each universe determines what is perceived and what constitutes valid knowledge, and the members of each society believe what is perceived and perceive what is believed.

□　　□　　□

A universe rises, flourishes, then declines in the course of time and is superseded by another. Its decline and fall occur because the society is assaulted by an alien culture, or startling new facts and ideas emerge, or old problems erupt and refuse to stay suppressed, or for no other reason than a steady change in the climate of opinion.

Unsurprisingly, within any moderately well-developed universe, the end of all knowledge seems to be in sight. A few things remain to be discovered, a few problems need to be solved, then everything at last will be crystal clear. Always, this or that topic of burning interest is said to be the last frontier. Pity the people of the future! What will they do when all knowledge has been discovered?

This oldest of human conceits, which confuses universe with Universe, is alive today as much as at any time in the past. We are afflicted with the hubris that denies our descendents the right to different and better knowledge. Throughout history devout people have felt convinced that their universe was the Universe, their mask the true face. Always, the universe in which one lives is thought to be the Universe. Then, within an ace of explaining everything worth explaining, the old and outworn universe dissolves in the ferment of social upheaval and transforms into a new and youthful universe full of challenge and expectation.

Cosmology is the study of universes. It is a prodigious enterprise, and naturally, cosmologists occupy themselves mainly with the study of the contemporary universe. One universe at a time is usually more than enough.

Yet undeniably we cannot understand the contemporary universe—and see it in historical perspective—without heeding the earlier universes from which it springs. Through the historian's eyes we see the past as a gallery of grand cosmic pictures, and we wonder, is our universe the final cosmic picture, have we arrived at last at the end of the gallery? We see the past as a procession of masks—masks of awesome grandeur—and we wonder, will the procession continue endlessly into the future?

And if there is no end in sight to the gallery of cosmic pictures, no end to the mockery of masks, what are we to make of the universe in which we now live?

More than 2,000 years ago, amid the social and intellectual revolutions of the Hellenic world, the Skeptics had their answer. Truth and justice are only what you care to make them. The Skeptics lacked confidence in an ultimate reality and disbelieved in the existence of the Universe. Their answer is the opposite of that presented in these pages.

□ □ □

The Masks of the Universe divides into three parts. Chapters in the first part deal with some universes of the past: the magic, mythic, geometric, medieval, infinite, and mechanistic universes; they form brief case studies of the cosmic beliefs in earlier societies, chosen for their interest and contribution to the modern physical universe. Our attention focuses less on the customary scenic details of history and more on the rise and fall of cosmic belief-systems.

I start with a speculative, probably fictional, account of the magic universe that supposedly arose hundreds of thousands of years ago when *Homo sapiens* began to acquire advanced linguistic skills. The magic universe—an animistic world actuated by psychic elements—developed into a living world, vibrant with ambient spirits motivated by thoughts and emotions mirroring the thoughts and emotions of human beings. Hosts of spirits pervaded the magic universe, and their behavior conformed to laws resembling the primitive social codes regulating the behavior of human beings.

I must point out that *magic* in this speculative account of the beliefs of early humans is used in a restricted sense; it does not denote the purely mystical and miraculous, but applies to whatever originates in the human mind as the assumed activating agents of the external world, such as spells, charms, apparitions, and spirits. I take the view that possibly nobody in the last 30,000 years has known quite what it was like to live fully immersed in the magic universe, and nobody since those earlier times has lived within a "primitive" human society in the true meaning of the word.

Across the span of hundreds of millennia the magic universe evolved into a magicomythic universe. The ambient spirits of the magic universe were amalgamated into nature spirits and assimilated into ascendent beings who personified increased unity and coordination in the phenomena of the external world. Many of the multivalent forms of the magicomythic universe survived until modern times in out-of-the-way places of the globe.

The mythic universe, reflecting the developing complexity of social life, arose some 10,000 to 20,000 years ago. (Mythic in this book is used to denote anything from one universe that does not fit naturally into another.) The mythic universe—mythic because its basic elements fail to fit naturally into the modern physical universe—was an enlarged universe ruled by remote and mighty gods who controlled and had created all that existed. This new and unified world view reached an advanced stage in the delta civilizations of Mesopotamia, Egypt, and India, and attained its most advanced expressions in the Zoroastrian and medieval universes.

The coordinated and unified mythic universe was purchased at a stiff price. The world of matter—of rocks and trees and clouds—became spiritless and dead. In an enlarged and transfigured world, riven by the dualities of good and evil, soul and flesh, free will and predestination, the timeless tales of the mythic universe tell of the tyranny of divine kingship, of incessant sacred wars commissioned by gods, of appeasement of the gods by human sacrifice, and of the massacre and enslavement of nations worshiping other gods.

We then witness the rise of science—the rejection of the gods as the sole agents of explication—and the development of the Ionian, Pythagorean, and Aristotelian world systems.

In the high and late Middle Ages the magisterial medieval universe dominates the skyline as perhaps the most satisfying and self-sufficient world system ever devised by the human mind. Here we see an age of devout wisdom, great scholarship, new science, and innovative technology. Indeed, its widespread technological revolution culminated in a style of life unique in the annals of history and laid the foundation of modern Western society.

Scholars in the high and late Middle Ages formulated notions that opened the way for the development of the Newtonian infinite universe. This universe rose to eminence in the Age of Reason in the eighteenth century (the century of progress), flourished in the nineteenth century (the century of evolution), and ushered in the modern physical universe that has overturned the mythic world of dead matter.

Chapters in the second part deal with the modern physical universe. I discuss those aspects on which our ideas have altered and are still changing. My intention is to stress what seems most interesting and to weave in the narrative strands from earlier themes. Beneath the surface of the physical universe lies mystery as intriguing as any found in the magic and mythic universes. Science reawakens the dead matter of the mythic universe with an inlay of vibrant activity, and the physical universe is now akin in many ways to the old magic universe. But the coruscating agents of explication dance more brilliantly and intricately than ever before. Much of science, I dare to say, consists of magic disciplined by a calculus of mythic laws.

In the third part I alight on miscellaneous topics of cosmological interest, starting with the witch universe. The witch universe arose in the late Middle Ages and terrified the Renaissance. It provides us with an interesting study of a mad universe on the rampage. Also, it illustrates my point that all universes are verifiable and falsifiable, and not just our physical world, as is often supposed.

Cosmology plucks fruit from all branches of learning and awakens a realization that around us in the modern world is another of the many universes, yet another of Keats's

> Magic casements, opening on the foam
> Of perilous seas, in faery lands forlorn.

Many and strange are "the universes that drift like bubbles in the foam upon the River of Time," writes Arthur Clarke in *The Wall of Darkness*.

The magic casements are high windows, revealing Scheherazadean vistas of theistic and mechanistic worlds: each world a universe constrained in scope by principles of containment, each constantly verified according to its terms of reference. In each we fail to portray ourselves fully as the conceivers of that universe, and no universe can ever be the Universe. Here are the perilous seas where human beings flounder amid the dualities of mind and matter, freewill and determinism, good and evil, and where they strive in the modern world to preserve the finer elements of the mythic universe.

□   □   □

That it might be possible to discover some of the general principles concerning universes has occurred to me in recent years. This thought is prompted by several questions.

Why are universes of various societies—Eskimo, Hindu, Polynesian, and the rest—so different, and what are their basic similarities?

Why do universes constantly change, is there perhaps a principle of natural selection whereby the "fittest" survive, and do they evolve because of developments in the human mind, or does the mind evolve in response to the need for more elaborate and better-organized universes?

Psychologists investigate the pathology of the mind; should not cosmologists in an analogous manner investigate the pathology of universes, seeking to understand why some—such as the witch universe of the Renaissance—are psychotic?

What determines the design of a universe; is it the Universe, God, fortuity, or the human mind?

[5]

The Universe and God are unknown; both are inconceivable and all-inclusive. Were Spinoza the philosopher and Goethe the poet right when they said the Universe and God are one and the same?

I have only gleanings and guesses to offer. What emerges, for me at least, is the overarching principle that distinguishes between our known universes and the unknown Universe.

☐   ☐   ☐

I have introduced the universes that are the cosmic belief-systems of societies. But individuals do not live entirely in the universe of their society, and I must also introduce *world pictures*, the multifarious worlds in which individuals live. Our world pictures are exceedingly complicated, and I shall approach them cautiously by way of a preamble.

In the spirit of the exact sciences we study the brain and observe it as an objective entity; we perform experiments and make measurements, and endeavor to disentangle its structures and their functions. We avoid confusing the inquiry by disavowing terms such as consciousness and soul that are objectively unmeasureable and physically meaningless. We stick to one outlook and the logic of one linguisitic system.

Information from outside the physical body impinges on the sense organs and is relayed by neurological pathways to the brain. The torrent of incident information in the form of light rays, sound waves, and other signals greatly exceeds what is needed and can be handled at any instant; the sense organs, pathways, and intermediate structures filter and regulate the incoming flow of information to the brain. The visual cortex, for instance, has interconnected clusters of cells responding selectively to movements and shapes, to edges and lines of particular orientation, thereby enhancing specific features in the visual field. The brain consists of ten or more billion neurons; each neuron has radiating fibers that link it to thousands of nearby and distant neurons, and the brain acts as an omniconnected network that receives, processes, stores, and interrelates information.

The brain may be viewed as a hierarchy of horizontally and vertically coordinated structures. At the highest levels in the cerebral cortex are arrays of interconnected neurons that account for short-term memory, the characteristics of individual behavior, and the peculiarly human ability whereby the brain studies the brain. Here, and at lower levels, are the soft-wired structures programmed by previous personal experience, language, and cultural inheritance. The hard-wired structures at much lower levels, linked with the glandular chemistry of emotional response, operate either automatically or semiautomatically in genetically determined ways.

Much of the brain is *terra incognita,* and we have yet to understand the physical basis of memory, sleep, and simple things like headaches. Colin Blakemore declares in his BBC Reith Lectures *The Mechanics of the Mind*, "The study of the brain is one of the last frontiers of human knowledge and of more immediate importance than understanding the infinity of space and the mystery of the atom." The study of the brain is undoubtedly of vital importance, for without knowledge of the way the brain works, how can we, who are brains according to this outlook, have confidence in the brain's reconstruction of the external world? Blakemore concludes his lectures with the words, "The brain struggling to understand the brain is society trying to understand itself."

Let us turn now to the other outlook expressed in the language of the subjective world. In this familiar mental world of thoughts and feelings and self-awareness we again must be circumspect and not confuse the analytic inquiry with borrowed expressions such as nerve impulses and glandular chemistry that lack operational meaning and rightly belong to the other outlook. We should also avoid such phrases as that used by Blakemore, claiming "our minds as products of our brains," which subordinate and tend even to eliminate the subjective outlook.

We may suppose that the psychic complexity of the human mind matches the physical complexity of the human brain, and there is thus an analogous hierarchy of mental structures. At the highest levels of the mind are realms of conscious awareness consisting of organized thoughts, where the mind contemplates such things as the nature of the brain and mind. As we descend to lower levels, cognitive processes become less conscious and more involuntary, concept-structures loom into view implicit in language, logic, and our workaday notions of space and time. Deeply rooted belief-systems, rich in emotive imagery, abound in this middle realm. At much lower levels we grope our way in the gloom of a subterranean realm shaped by primate evolution and hominoidal experience.

The objective brain interacting with the external physical world and the subjective mind interrelating its realms of psychic experience are alternate scenic backdrops to the same play. In our everyday life as physicians, psychologists, and ordinary persons we use the complementary scenarios continually, switching from one to the other ("How are you feeling today?" Now let me see what is bothering you"), blending them into a unity of overlapping outlooks. Together they enrich our world, and normally we are unaware of their radical dissimilarity.

We study the brain interacting with the physical world of which it is a part, and we have a reasonably systematic description, provided we leave aside confusing psychic terminology. We study the mind and its conscious and unconscious realms, and again we have a reasonably systematic description, provided we leave aside confusing physical terminology. In each case the functional elements are moderately clear when we adhere consistently to a single outlook. But neither outlook by itself accounts adequately for the whole of human experience. Joining the two lies a tangled terminological web that supposedly relates the objective and subjective worlds in which we live. No philosopher or scientist has yet succeeded in disentangling this web. The best and perhaps only thing possible in the context of the modern physical universe is to accept the complementary outlooks as inherent in the human descriptive process.

The events of the external world lack a direct and simple one-to-one correspondence with the internal events of the brain and mind. The external world is mapped within the brain (and the mind) in a highly convoluted manner; each mapping forms a unique configuration (a world picture) peculiar to one person, and the features common to a class of configurations (a universe) are peculiar to a society.

We reconstruct our experiences by deconvoluting these configurations of the brain-mind and create two worlds, one impersonal (a universe), referred to as objective, the other personal (a world picture), referred to as subjective.

☐　☐　☐

The members of a society share their thoughts and experiences; they have a common scheme of concepts that constitutes the universe of their society. This universe possesses numerous sets of supplementary and complementary relations; when an experience cannot be explained by one set of relations, it is explained by either another set or a combination of sets. Subordinate communities, such as kinship groups, have their own particular interests and beliefs, and the embracing universe succeeds in accommodating all communities within a unified society.

But individuals cannot live entirely in the impersonal universe of their society. Each has a private world of knowledge and emotion, of thoughts, hopes, and dreams. These private worlds, which I call world pictures, are of interest to psychologists and others working in the cognitive sciences. A world picture faces outward and inward; it looks outward to a societal universe of common knowledge and inward to a personal world of emotive imagery; it is a particular configuration of the mind-brain, unique to one person.

We follow a strange path: on one side societies and their universes, on the other individuals and their world pictures. Confronting us is the spectacle of universes and their world pictures, of each universe unifying a cluster of world pictures into a social unity of individuals.

Let me hasten to say that this book is about universes and has little to contribute on the subject of world pictures. Our world pictures are much too varied and complex, far too disorganized, in fact, to play a footlight role in the burlesque of cosmology, and they are offstage among the audience. World pictures are the repositories of mythic elements from earlier universes. Most of the rational lucidity they possess is already woven into the fabric of the contemporary universe; what remains was once rational or will become rational within the fabric of other universes. As cosmologists we disdain our disorganized world pictures and regard them as little more than crazy belief-systems. Yet they are the wellsprings of universes. The deep structure of the mind, implicit in world pictures and partially explicit in universes, awaits further elucidation within the fabric of subsequent universes.

A person possesses an elaborate belief-system that Seymour Epstein in *The Self-Concept Revisited* refers to as a "personal theory of reality." According to Epstein, a personal theory of reality assimilates and formats an individual's experiences, maintains self-esteem, and creates a favorable pleasure-pain balance over the foreseeable future. A personal theory of reality is like a world picture if we interpret theory to mean "view," as in the original Greek sense. But a world picture is more than a belief-system, more than a personal theory of reality, for it comprises the whole person, of "me" and "not me," and includes the external world from the perspective of the individual. You are a world picture, and I am a world picture.

World pictures invest the societal universe with personal meaning and value judgment, color it with emotion and sensuality, clothe it with beauty and ugliness; they sustain us with fanciful hopes for the future and reassuring memories of the past. The lushness of our personal world pictures compensates for the aridity of the impersonal universe.

□   □   □

Often we pretend not to live in the universe, knowing that we pretend. We alternate between no pretense when we live in the "real" world of our society, and double pretense when we pretend to live in a pretended world and "all that we see or seem is but a dream within a dream." It is the natural way a sane person lives. Into our counterfeit worlds of fantasy we withdraw when the reality of the universe becomes too grim. On returning, we put down the novel, turn off the television, or come home from the play, feeling refreshed and entertained, knowing that we have lived in a counterfeit world.

But those individuals lost and tragically betrayed by the universe, who cannot alternate between no pretense and double pretense, who find sanctuary in an inner world of pretense and never realize the extent of their pretense, they are the insane. A world picture becomes psychotic when it disengages from its societal universe and substitutes a private world of fantasy believed to be real.

It is the old question: who is insane and who is sane? The one secure in an isolated world picture, or the other exposed to a societal universe and its incessant conflicts? We say the one who sojourns in an estranged and secret world of solace is insane, for that person has withdrawn from society into a world picture shattered by the universe.

Yet sanity can be overdone. The entirely sane person, if such exists, is only half alive. The excitement of living comes from the conflict of world pictures and a universe. We seek always to impose our will and shape a universe to our desire. The struggle is fierce, the casualty rate high, and the badly wounded escape from the struggle into worlds of madness. Those left on the battlefield engaged in the war of the worlds, whether in cloister, study, office, workshop, farm, or counterculture community, are the robust ones who bear the brunt and yet have the thrill of high adventure. Great sages, visionary prophets, creative thinkers, and imaginative artists have fought their universes and triumphantly imposed their will.

But what of the universes that betray not just a few but many of the world pictures of their societies? These are insane universes created by the tyranny of aberrant world pictures. In the annals of history the mad universes are many. We need only mention the witch universe of the Renaissance, the universes of societies engaged in bitter religious and civil wars that sicken everyone and are sustained by leaders with sick minds, and, more recently, the twentieth-century variants of the physical universe in thralldom to the reckless mania of totalitarian ideology. A mad universe oppresses and destroys the world pictures of individuals, imposes termite-like uniformity, exalts the authority of society, and usually but not always is short-lived.

A mad world picture isolates itself from the real world as determined by its societal universe. A mad universe eliminates rather than unifies the diversity of individual world pictures in its society.

While I write, hosts of Wordsworth's golden daffodils are "fluttering and dancing in the breeze." You and I live in the world out there of hills, lakes, and daffodils, of multitudes of things and torrents of events. And the grand picture we all share in Western society, we all believe in more or less, is the physical universe.

Many among us understand little about the structure of atoms, cells, and stars, and some may even dislike the physical universe. But we drive automobiles while listening to the radio, fly in planes to distant lands, watch television, rely on computers, trust in modern medicine, and use electricity in a myriad ways. We may not understand, and may not like, yet we depend on and to a great extent believe in the physical universe.

The people in the past, as we know, had other outlooks and ways of viewing their world and lived in, or thought they lived in, a different kind of universe. The universes of the past—Babylonian, Pythagorean, Aristotelian, Newtonian, to name a few—were all different, and none was like the modern physical universe. In the Babylonian universe the flowers danced and fluttered in the breeze, the Sun rose and set, the Moon waxed and waned, the jeweled lights in the night sky traveled the heavens, and a rock was a rock, and a tree a tree. But the nature of these things and their meaning was other than what we now think. The lifestyles and modes of thought of the Babylonians, so unlike our own, were in harmony with the Babylonian universe.

Many persons of today when asked would unhesitatingly agree that the out-of-date and discarded universes of the past were all mistaken in their detailed and general view of things. But—and here is the rub—it does not take much imagination to realize that the people in the past believed in their universes just as firmly as we tend to believe in the modern physical universe.

A person of today living in our society might dislike the modern physical universe, and a Babylonian might have disliked the Babylonian universe, but this is an irrelevant issue and quite unimportant. To be a member of a society, one thinks in terms of the universe of that society, quite apart from whether it is liked or disliked.

The people in the past believed in their universes. Here is a fact we tend not to dwell upon because of its disconcerting implications. We see the people in the past believing in the truth of their universes, and because they were mistaken, might not we also be a little mistaken? And if a little, well then, why not a lot, like all the rest? We dismiss the thought on the grounds that the people in the past were ignorant.

But the thought persists. If what happened in the past is a guide to what will happen in the future, then conceivably our modern beliefs are also greatly mistaken, and perhaps one day a new universe will arise, grander than the modern physical universe. Those living in the future will look back in history and see our twentieth-century universe as out-of-date and mistaken like all the rest.

"It seemed to me that no more fitting preface could be put before a Journal, which aims to mirror the progress of that fashioning by Nature of a picture of herself, in the mind of man, which we call the progress of Science," wrote Thomas Huxley in 1869 for the first issue of the science journal *Nature*. I have discussed in this introductory chapter the plurality of universes, and in effect paraphrased Huxley by saying that the Universe fashions pictures of itself in the mind of man, which we call universes. It was not my intention, nor I imagine Huxley's, to give the impression that a universe is a fancy-free invention, as put forth in *Romeo and Juliet,*

> *Begot of nothing but vain fantasy*
> *Which is as thin of substance as the air*
> *And more inconstant than the wind.*

We are not like gods free to spin firmaments and baseless fabrics out of thin air. The worlds in which we live as sane individuals are not the stuff of dreams, and, from *The Tempest*, our universes of

> *. . . cloud-capped towers, the gorgeous palaces,*
> *The solemn temple, the great globe itself,*

are not the insubstantial pageant of a playwright.

More than once the reader may have wondered how universes can come and go, tra la, like flowers that bloom in the spring. Surely there is only one universe—or Universe—whatever name one wishes to use, and are we not acquainted with it? Surely everybody is aware of the same things—chairs and tables, rocks and trees, the Sun and the Moon—and thinks about them in much the same way? Those things out there must in some sense be real and permanent. Surely the advance of knowledge consists of uncovering little by little all the details, of discovering step by step all the correct explanations?

George Berkeley, Irish philosopher and bishop in the early eighteenth century, argued that only our mental experiences are real, minds and God alone exist, and the external world is an illusion emanating from God. James Boswell in his biography of Samuel Johnson wrote, "We stood talking for some time together of Bishop Berkeley's ingenious sophistry to prove the non-existence of matter. ... I shall always remember the alacrity with which Johnson answered, striking his foot with mighty force against a large stone, till he rebounded from it—'I refute it thus'." Few would disagree with Johnson's impressive demonstration of the reality of the external world. Yet facts are rarely as concrete as the stone Johnson attacked. "Where," asks Morris Kline in *Mathematics in Western Culture*, "is the good, old-fashioned solid matter that obeys precise, compelling mathematical laws? The stone that Dr. Johnson once kicked to demonstrate the reality of matter has become dissipated in a diffuse distribution of mathematical probabilities."

The reader in a note of desperation might ask, Where, then, are the concrete facts of reality? Notwithstanding the hardness of Dr. Johnson's stone, the straight answer is that apart from the immediate elements of personal experience, we are quite unable to ascertain with certainty and finality what are the abiding concrete facts.

The outer and inner worlds of experience are undoubtedly parts or aspects of the Universe, but how we describe and explain our experiences is no more than a universe peculiar to a society at a particular time. Each universe, moreover, evolves because of the struggle from within of world pictures in rebellion, and because of the struggle from without of universes in competition. The first part of this book dwells on these struggles from within and without.

The realization that universes are impermanent conceptual schemes is the outcome of trends of thought in the study of the history of cosmology. This disturbing aspect of cosmology is rarely stressed and might therefore come as a shock to some readers. Without much thought we tend to regard the universes of other and earlier societies as pathetically unreal in comparison with our own. It is disconcerting to be told that our modern physical universe is just the latest model that almost certainly in the future will be discarded and replaced with another and possibly more resplendent model.

This exercise in the historical imagination extends the scope of our thinking, and we realize how pathetic perhaps many of our present cherished cosmic beliefs will seem to those in a future age living in a different climate of opinion and having a different mental temper.

At present in our modern universe we are confronted with the problem of rocks and trees *and* the observer, of the brain reconstructing a world of rocks and trees *and* the brain. Perhaps the next universe will make clear these issues by enabling us to view them within a wider conceptual scheme. We dare not omit from reality the convoluted imagery of the inner world, and already we sail the perilous seas where the universes in which we live are recognized as reconstructions by the mind-brain.

◻ ◻ ◻

Arthur Eddington, a scientist who leaned toward philosophy and wrote fascinating books that attracted youths of my time into physics, once said, "We have found a strange footprint on the shores of the unknown. We have devised profound theories, one after another, to account for its origin. At last we have succeeded in reconstructing the creature that made the footprint. And lo! it is our own."

Most of us would agree with Eddington that the properties of the mind-brain are in some way imprinted in the physical universe. When, however, Eddington adds, "The mind has but recovered from nature what the mind put into nature," he ventures too far, further than seems warranted. If we do not understand ourselves, how can we hope to recover what we do not understand and cannot recognize? We are a part or an aspect of "nature," involved in the process of nature observing and understanding itself, and this is a more complex situation than that considered by Eddington.

Albert Einstein once said, "The most incomprehensible thing about the universe is that it is comprehensible." To this one might add, The most comprehensible thing about the Universe is that it is incomprehensible.

We cannot doubt the existence of an ultimate reality. It is the Universe forever masked. We are a part or an aspect of it, and the masks figured by us are the Universe observing and understanding itself from a human point of view. When we doubt the Universe we doubt ourselves. The Universe thinks, therefore it is.

Part 1

Worlds
in
the
Making

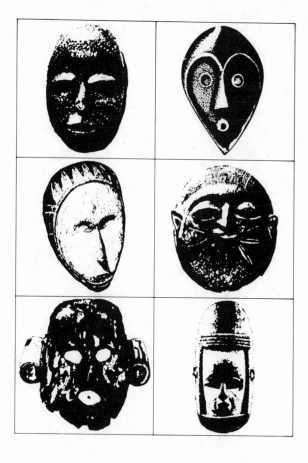

The Magic Universe

"History is only a pack of tricks we play on the dead," said Voltaire. By scanning history, peering back into prehistory, we seek the ancestral incunabula; with meddlesome curiosity we turn over stones, dig up bones, and expect the dead of long ago to forgive the tricks we play.

At least we have learned not to portray the early human beings as shambling Nibelungs, as Hobbesian ogres, "solitary, poor, nasty, brutish, and short." Doubtless the forgotten people of the distant past were thoughtful beings, with a spring in their stride and light in their eyes, who ornamented their bodies, bedecked their dead with flowers, danced, sang, laughed, cried, and, like us, had their joys and sorrows. They lacked our knowledge, yet had instead their own, perhaps more than we realize.

Very little is known of the early human beings who lived hundreds of thousands of years ago. Their style of life was certainly primitive by our standards, and probably even by the standards of the hunting-gathering African Bushmen and Australian Aborigines. Other than a miscellany of skulls and skeletal remains, tool kits, artifacts, and evidence of diet, we have precious little information on how the early human beings lived, and none whatever on how they thought. But we know their brains were as large as our own, and we may safely assume they used them.

The universe in which the early human beings lived, or thought they lived, is lost forever, and all our speculative reconstructions are probably in error. My guess is the following.

□   □   □

Imagine a group of hairless and thin-skinned striding primates, encumbered with juveniles taking a decade to reach maturity and with elders needing special care. This picture of early human beings wandering on open savannas, along seashores, and through woodland forests prompts us to wonder how they could survive, when the animals around them were fleet-footed, protected by fur, and armed with sharp claws, horns, long teeth, and tusks.

True, in their skillful hands the art of using tools had long before climaxed in the technology of toolmaking. "Man is a toolmaking animal," said Benjamin Franklin, and marked progress in toolmaking was undoubtedly associated with the development of the human brain-mind. Toolmaking provided the weaponry that compensated for a defenseless physiology and enabled the early human beings to survive. Yet we go much too far if we say that toolmaking accounted for the breakthrough to sophisticated brains and minds. The manufacture of carrying bags (one of the greatest of all inventions), the employment of fire (at least half a million years ago), and the industry of toolmaking provide evidence of intelligence, yet by themselves cannot be regarded as the primary cause of intelligence.

Our vignette of primates equipped with carrying bags, fire, tools, and weapons is incomplete. It overlooks the supreme fact that they are chattering together. The breakthrough to sophisticated brain-minds had started long before, when our remote forebears developed the power of speech and acquired linguistic skills. Language was the potent means of organizing and unifying social groups capable of living and roving in unsecluded environments.

Three million years ago the *Australopithecus* hominids of South Africa had a cranial capacity of 400 to 500 milliliters, already larger than that of chimpanzees; a million years later *Homo habilis* had a brain volume of 600 to 700 milliliters; the rate of increase was rapid, and a million years ago *Homo erectus* had increased to 900 to 1100; modern human beings soon emerged with an average cranial capacity of 1450 milliliters. The differences of significance between human beings and apes are brain size and language ability, and not unreasonably we must suppose that the panoply of mental processes accompanying linguistic ability is linked with the complexity and average size of the hominid brain.

Apes communicate with sound and gestures, and their signals to one another enable them to live as groups in sequestered environments. But language with its array of symbolic structures and command of articulate and fluent speech is vastly more than a repertory of cries and gestures. "Language is a . . . noninstinctive method of communicating ideas, emotions, and desires by means of a system of voluntarily produced symbols. These symbols are, in the first instance, auditory and are produced by the so-called 'organs of speech'," wrote Edward Sapir, a pioneer of modern linquistics, in his popular book *Language*.

We have a picture of a tightly knit group of jabbering individuals who share their experiences. They live on a mixed diet, hunting and gathering, and it is a fair complaint, remarks Williams Howells in *Evolution of the Genus Homo*, "that man the hunter has been extolled at the expense of woman the gatherer." Men and women, then as now, had equivalent opportunities for the exercise of intelligence and courage. And much to our surprise, the early human beings did not live in constant fear of a hostile world. They consulted together, formulated plans, acted on command as a unit, referred to a cultural memory of effective strategies, and employed devastating tactics of alternating offense and defense. With language was forged the mightiest weapon on Earth. Men and women are talking animals.

Like a bouyant force, language lifted intelligence to higher levels. Articulate thoughts interlaced facts within a widened expanse of memory, and greater intelligence made possible more elevated forms of linquistic expression. Intelligent minds were naturally selected, for whoever could not find the apposite words, comprehend and obey the voice of command, recall the effective strategy, or respond with the efficient tactic had little chance of surviving. One might say, "man is a heroic animal," for the earliest hominids and human beings trod a precarious path of immense challenge. Perhaps many hominid species started and failed, and some retreated back into their sequestered worlds. Chimpanzees, it has been suggested, are dehumanized hominids who withdrew from the challenge.

We lack an acceptable general method of measuring intelligence. But intelligence tests are not an invention of the modern age, for Nature once had her own (perhaps still has), and dispensed judgment in her usual forthright fashion. Candidates with low scores were eliminated, and we are the beneficiaries of the weeding-out process.

Children take a comparatively long time to reach maturity, and we have evolved in that way because many years are needed to learn the language and cultural wealth. This fact alone indicates how great was the knowledge our remote ancestors handed on to their offspring. Rapid physical growth would have had a distinct disadvantage when prolonged care by the family was needed during the helpless years of learning.

Those groups careless and impatient with their young did not survive for long. In the hunting-gathering groups the young were taught the language and initiated into the laws and cosmic truths. The old were cherished as wise leaders, as guardians of the cultural heritage, and the groups who discarded their old or were indifferent to their care also did not survive for long.

The lifestyles of the natives of Australia and South America, the Shoshones of North America, the Pygmies of the Congo Valley, and the Bushmen of the Kalahari Desert offer us clues concerning the modes of life of our early forebears, but the clues are slender and possibly even misleading.

☐   ☐   ☐

Anthropologists have speculated on how the people of long ago viewed their world. In *The Intellectual Adventure of Ancient Man*, Henri and Groenewegen Frankfort, John Wilson, and Thorkild Jacobsen suggest that the world appeared to primitive humans "as neither inanimate nor empty but redundant with life." Everything was living, they argue, and

> ... life had individuality, in man and beast and plant, and in every phenomenon which confronts him—the thunderclap, the sudden shadow, the eerie and unknown clearing in the wood, the stone which suddenly hurts him when he stumbles while on a hunting trip. Any phenomenon may at any time face him, not as "It," but as "Thou." In this confrontation, "Thou" is not contemplated with intellectual detachment; it is experienced as life confronting life, involving every faculty of man in a reciprocal relationship.

The early people lived in a universe animated by life. The simplest and most effective means available for understanding how organic and inorganic things behaved was to know that they were living just like the early people. It was not assumed but instinctively known that all things lived. Distinctions between animate and inanimate states were no more than differences between being awake and asleep. Whatever demanded and received attention, and seemed significant enough to be distinguished by name, had its individual life and mode of behavior.

In the opening act, hundreds of thousands of years before the present, it was perhaps little more than a behavioristic sort of world in which numerous things had their identifying names and distinguishing patterns of behavior. It was a living world animated with rudimentary forms of life, in which each lifeform seemed an elementary unity—nothing more than an undifferentiated whole—with no distinction made between outer physical body and inner mental being.

The recognition of one's own inner mentality of feelings and thoughts, of one's subjective self as distinct from the physical body and objective nonself, and the ensuing attribution of this mentality to other persons, came gradually and automatically, presumably as a consequence of growing intelligence and social influence. The early human beings uncovered the depths of personality and enlarged their world by conceding to one another an inner mentality, a state of conscious being, expressed in a wealth of linguistic terms. Each person knew that his or her own impulses and emotions were similar to those of other members of the social group. Greater intimacy and interdependence in family and social living followed. Quite probably at this stage man the hunter and woman the gatherer became mutually supporting within a family unit.

Inevitably, the projection of the inner self into other human beings widened to include beasts and plants. Then, in the course of time, all significant things acquired humanlike personalities. Only in this way could anything be explained. At last we stand at the threshold of the magic universe.

Human desires and impulses animated everything, and the world was alive in every conceivable sense. The mentality of the lifeforms of the magic universe mirrored the mentality of the human beings who lived in it. It was a looking-glass universe capable of explaining all known phenomena and the entire range of human experience. A world of unlimited richness had emerged, in which the mind was constantly strained to its utmost limits, a charismatic world that conferred on its occupants an ability to transcend all other human beings still living in simpler worlds. Out of a human population amounting perhaps to no more than a million, only a few groups, consisting of hundreds of individuals, crossed the threshold into the magic universe. Their newfound world of imaginative power gave them superior ability to survive.

The word magic is widely and loosely used in many contexts. Here I have honed it down to mean nothing more than the human mind made explicit in the external world. If you believe in angels, fairies, Santa Claus, demons, the evil eye, and other anthropopsychic agents active in the external world, accounting for what is observed, then you live in a sort of magic or thaumaturgic universe, but nothing like the world of the early people as I shall try to show. Magic is the mind interfacing with a world in which the creations of the mind are the explicit activating agents. Science activates the world with subtle impersonal principles, and it seems safe to say that science derives from magic.

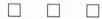

At some stage, impossible to guess when, but still very long ago, the activating psychic complexities of the magic universe evolved to the stage at which many attained a kind of independent reality. By independent reality I mean the inner psychic being became detached from or tenuously connected to its outer physical form. In most cases this psychic being—or spirit—was capable of enduring after the dissolution of the physical body. Many of these spirits had forms other than their ordinary bodies and frequently were so tenuously connected to their physical counterparts as to be virtually disembodied.

Animism is the firmly entrenched knowledge that the things observed have their indwelling or associated spirits. The animated universe had deepened into an animistic universe densely populated with spirits. I am sure that our present-day fine distinction between spirit and soul would not have been understood by the early people.

Plausible reasons account for why the spirits came into existence. Reluctance to accept death as natural and inevitable perhaps inclined the early people to think that the inner mentality never died and was therefore distinct from the mortal body. In the dreamstate the inner mentality seemed capable of leaving the body and traveling elsewhere. Possibly, the early people had a view of time in which events of the past appeared to coexist with events of the present, and nothing died, but only changed from a corporeal to an incorporeal state. The simplest of all reasons, however, is that the advent of spirits was no more than an inevitable consequence of the initial animated nature of the universe of primitive human beings.

By means of spirits the early people gained deeper understanding and greater control of the phenomena of their world, and in the ceaseless competition of universes, cultural selection processes took care of the rest and facilitated differential survival.

Language exploded, expanding in scope to encompass the concepts of spirits, and human beings were able to refer to abstractions and to think of the properties of general phenomena. Rivers, lakes, mountains, valleys, clearings in woods, the camp or domestic cave became identified by their ambient psychic beings, and the earth, sky, wind, water, and fire were invested with diffuse nature spirits.

The magic universe, pulsating with spirits of every kind imaginable, reflected and magnified the emotions and thoughts of human beings. A veneer of physical forms overlaid a world of benevolent, indifferent, and malignant spirits that resonated with the inner world of each person and amplified all mental experience. Dreams, entrancements, and altered states of awareness contributed to its kaleidoscopic imagery.

It was a vibrant universe awakened each day by the Sun spirit and mourned each night by the Moon spirit. A universe of starlike campfires across the night sky; of chromatic sky spirits manifesting as rainbows, sunsets, and northern lights; of mighty earth spirits rumbling beneath the ground and spewing forth from volcanoes; of flittering little folk dwelling in secret places and stealing lost children. A magic universe haunted by the unborn and dead forever calling. Words cannot recall nor the mind recapture the extreme vividity of its imagery. On windy nights the trees awoke, swaying their contorted branches, conspiring in sibilant voices, creating abject terror among the huddled human beings and their familiar spirits.

The sudden noise, the fallen tree, the shaft of light piercing the forest gloom, the rising river, the lowering sky, the hurtful stone, and each incident of every day was the natural consequence of incarnate spirits pursuing their various interests. It was a numinous universe of the kind fleetingly glimpsed by young children in spine-tingling fairy tales.

The more the early people spiritualized the magic universe, the more humanlike they became themselves.

☐ ☐ ☐

Wherever there is a society we find a universe. A society of intelligent members, sharing thoughts and experiences, is the necessary and sufficient condition for the existence of a rational universe. A particular universe might not be rational by our standards, or those of other societies, but is always rational by the standards of the members of its own society.

The magic universe (or whatever one wishes to call it) was made rational by gauging the characters and moods of its activating psychic beings. It was not in any way illogical or prelogical but fully logical in accordance with its own principles. We must put aside those hoary tales telling us that primitive human beings lived in an irrational world in which little or nothing could be anticipated because of the capriciousness of the spirits. On the contrary, in accordance with its principles, the magic universe was fully logical, completely rational, and abounded with predictable phenomena. We would be quite wrong to suppose that every incident of significance always preceded rather than followed the nodding of wise heads.

The early people, regulated by their social customs and traditions, were no more capricious in behavior than we are today. Groups would have lost their internal cohesion and disintegrated if composed of persons of unpredictable and irresponsible behavior. Humanfolk and spiritfolk reflected each other; the spiritfolk, regulated also by customs and traditions, were no more capricious than the humanfolk. Human beings predicted the behavior of spirits to the extent that they predicted the behavior of one another.

Members of a primitive social group conformed more or less to the conventions of their society. A rebellious member could be coaxed by soothing words, loved by concerned kinfolk, shown in what way he or she stood to gain by pleasing others, shamed by indignation when the rules were broken, and occasionally coerced by dire threats. The spirits also conformed more or less to known rules and could be coaxed, loved, bribed, shamed, and even coerced into conformity. The aid of benign spirits was solicited, and the venom of malign spirits mitigated. And by offering gifts and performing helpful and pleasing tasks the human beings influenced the spirits in the same way and to the same extent that they habitually influenced one another.

By such means the early people gained control over their world and predicted many if not most of its events. The lowering sky gave warning of the imminence of storm spirits, and the forewarned people took shelter. A child while running for shelter with its mother might trip over a stone, and after the mother had scolded the hurtful stone, the child was never again tripped by the same stone. Everywhere the spirits displayed signs that made clear their moods and intentions, and the people read the signs and acted accordingly. By coaxing, loving, bribing, shaming, coercing the spirits, and by interacting with them in multiple ways, the people influenced and gained control over their world.

The magic universe was a relatively simple world of relatively simple societies consisting of life confronting life. What might seem to us a pitifully ineffectual world, on the contrary, seemed to the early people fully effectual. Quite possibly they had more understanding and control of their simple world than we individually have of our own complex world. How many persons of today understand the inner workings of an automobile, how a jet plane flies, and how to repair a television set? Yet these are among the commonest things around us. The early people not only saw their world as comprehensible, but they also influenced and controlled it, which is much more than can be said of most persons living today in the physical universe. I am inclined to think that of all universes, the magic universe was internally the most rational and lucid, and every subsequent development has been purchased at the cost of added mystery and perplexity.

☐   ☐   ☐

"Possessed, pervaded, and crowded with spiritual beings," said the Victorian anthropologist Edward Tylor, referring to the world of primitive man. In his *Primitive Culture* of 1871 he proposed the theory of animism and conjectured that animistic beliefs were invented by "the ancient savage philosophers." Theories of how the early people thought consist of little more than wild guesswork, and if perchance animism is the correct theory, as we have assumed, it seems most improbable that animism originated as a contrived philosophical invention in the manner proposed by Tylor. More likely it came naturally and accompanied the development of language.

Any attempt to understand primitive human thought is confronted from the outset by a daunting problem. To uncover primitive modes of thought we must first strip away from our minds all mental processes and thought habits that have since developed, and yet owing to our ignorance, we are never quite sure how much must be stripped away. The problem seems insoluble. The thoughts we ascribe to the early people in our own minds were not necessarily in theirs.

"I shall invite my readers," writes the famed Branislaw Malinowski, "to step outside the closed study of the theorist into the open air of the anthropological field." We buy our tickets and accompany Malinowski to the Trobriand Islands of Melanesia. There, on these islands, as described in *Magic, Science and Religion,* we find mana (a generalized spirit), totemism, fetishes, charms, shamanism, sorcery, and cults of vegetation and fertility are conspicuous, and primitive animism plays little or no role in the lives of the natives. The Trobriand Islanders work in their gardens and fish from their canoes using traditional techniques, drawing on a large body of empirical knowledge, in all of which supernatural beliefs are unobtrusive, apart from the shaman's indispensable ritual of occasionally blessing the gardens and canoes. Supernatural beliefs are visibly active in ceremonies and collective rites and in preparation for voyages and are of paramount importance on hazardous occasions when the natives fish in dangerous waters or there is fear—as in time of war— of dire misfortune. Supernatural beliefs hang in the background like a tapestry weaving together the threads of mortal and immortal life, customs and empirical procedures lie in the foreground regulating the affairs of everyday life. This is not the magic universe. Divested of idiosyncratic detail, and in schematic outline, it looks not entirely unlike the universe of many Europeans in the last few centuries.

Pure and simple animism is now in a cloud of doubt, because nowhere can it be found in the anthropological field. Anthropologists in their field investigations have not discovered anything so primitive, and the theory of animism appears to be incapable of explaining the sophisticated belief-systems of recent and present-day "primitive" societies.

The word primitive, denoting what is earliest or among the first, has much the same meaning as primary and applies to whatever is original. It is a seductive word, frequently misused, which confers a deceptive aura of simplicity. Call a thing primitive and the battle of explaining it is half-won. We label out-of-the-way people primitive when their life-styles and belief-systems are other than our own. The word is often a misnomer that leads us much astray. It divides us from the so-called primitives, placing them on the far side of a wide gulf of time, contrary to the fact they now are living and as intelligent as ourselves.

One might justifiably question whether in historical times any truly primitive society has existed. The societies familiar to us look much too sophisticated to be dubbed primitive. Their members have brains similar to and beliefs as complex as those of Europeans and members of other civilized societies who live in cities. (Civilized is also a much misused word but not worth berating.)

The assumption that recent and modern primitives fail to use their brains fully is self-serving and untenable. The human brain-mind, taxed to its limits continually, evolved to its present complexity under living conditions far more rudimentary than those of the so-called primitives of our time.

The languages and thought-habits of modern natives are as complicated as our own. Of course, we are proud that the opulent physical universe of Western society is more elaborate and more highly contrived than the ramshackle magicomythic universes of preliterate societies; after all, it is a cornucopian universe that makes possible the ships, planes, radio sets, cameras, tape recorders, computers, statistical methods of analysis, and knives and needles to bribe the natives, which are so necessary in anthropological investigations. But the personal and jerry-built world pictures of the investigators are no more rational and complex than the world pictures of the natives they study.

The otiose assumption that our society has evolved from societies identical with those now existing and labeled primitive is equivalent to the assumption that we have evolved from apes identical with those now living. We and apes have diverged over a great period of time from primitive hominoidal stocks, and similarly the societies covering the globe have diverged over long periods of time from primitive social origins. This may explain why primitive animism cannot now be found (any more than Neanderthals and earlier hominids) and why the magic universe has entirely vanished. Backsliding societies are not necessarily regressing to the original primitive state; more probably they are sliding down a different route.

The hypothesis of primitive animism possesses several advantages. Animism was a universal method of elucidation much simpler than any now existing; it originated naturally and automatically, without need of conscious and deliberate invention, offered a self-consistent rational scheme that provided ample scope for development, possessed adequate predictive power, and gave to the early people effective control over their world.

I shall remain in my study and with regret decline Malinowski's kind invitation. Unfortunately, the specimens of interest no longer exist.

☐   ☐   ☐

The magic universe slowly evolved and lost its original simplicity. An accumulation of knowledge in depth and extent at last culminated in a considerably changed outlook some tens of millennia ago. Spirits of power emerged.

Hitherto the customs and relationships of the spiritfolk had reflected little more than the customs and relationships of the humanfolk. The one side mirrored the other. Spirits were classified into species of bear, wolf, owl, ..., of plants, and into other kindred groupings. Each class of spirit had its lifestyles, social relations, and traditions recounted by people in an encyclopedic body of orally transmitted legends and lores. As human societies evolved and became more intricately organized, so with the spirit societies, and again the one side mirrored the other. But the mirror began to distort and magnify the spirit images. With more knowledge came a dawning awareness of the vastness and complexity of nature and a realization that the beings responsible for animating the universe were greatly superior to both ordinary humanfolk and simple spiritfolk. Step by step the magic universe evolved into a magicomythic universe. Among the possible offshoots of the magic universe I shall follow a single broad course.

The spirit taxonomy acquired hierarchical orders of graduated authority dominated by awesome representatives, veritable godlings, who possessed the requisite power to control the animations of nature. On one side of the mirror stood human beings and their societies; on the other side of the mirror arose Brobdingnagian spirits, no longer simple and humanlike, but complex and superhuman. The distortion in the mirror was this: on one side remained human beings confronting the animations of nature, and on the other side emerged exalted spirits of superhuman magnitude who reflected less and less the character of individual human beings, and more and more the character and collective power of human societies and their officialdoms.

Spirits of power intruded, and the magic of individual things waned. The little spirits, who once had activated everything in a haphazard fashion, or so it now seemed, who needed to be constantly watched and cajoled into compliance, lost their credibility and withered away. In effect, they were withdrawn and combined into greater, better-organized, more unified spirits. These greater spirits—exalted psychic images of social magnitude—with abilities and perceptions never granted to simple human beings and spirits, now orchestrated the phenomena of the world, and provided human beings with a more advanced understanding of the vastness and complexity of nature.

Informal and spontaneous speech with ambient spirits declined, and invocation of great godlings and universal nature spirits became the arcane preserve of privileged officials. The magic apparatus of minute by minute explication was dismantled, and the mythic machinery of cosmic order and unity installed. Ceremonial food offerings and sacrifices placated the fearsome godlings; invocations and incantations guaranteed the maintenance of food supplies; institutional worship sustained the rhythm of the seasons. Totemism and its clan dependence on exclusive relations with spirit-species had arrived; to hunt and kill required permission not from the animal itself, as hitherto, but from the collective spirit of its species, and this permission was best obtained by the clan members having totemic identity with the animal. The cave paintings of the Paleolithic are conceivably symbolic of totemic spirits to whom appeal was mandatory to guarantee success in hunting.

This kind of universe, regulated by superhuman spirits and half way between the magic and mythic universes, is what we seem to see in the recorded accounts of the isolated societies of Australia and of North and South America. The magic universe had evolved and branched into the many worlds of the intermediate magicomythic universe, each world possessing exalted spirits that reflected not only the characteristics of human beings, but also those of the institutions of their societies.

The magicomythic universe brought advanced understanding and gave to human beings, through the intercession of exalted spirits, greater control over their world.

□ □ □

By comparison with the magic universe, the magicomythic universe was a unified, far-seeing cosmos that sustained people in their various quests and inspired and nerved them to undertake building projects and explorations by land and sea previously unthinkable. The primitive spirits of the magic universe had never shown much interest in distant places, but not so the godlings of the magicomythic universe, who ruled far and wide in space and time. Each magicomythic world sustained the members of its society with a deep awareness of cosmic identity.

The many worlds of the magicomythic universe collided and erupted in turmoil.

The jealous godlings who ruled the divergent magicomythic worlds, and were contemptuous and intolerant of rival authority, drove their societies into open conflict. When overwhelmed by conquest, only then could one society accept the notion that its godlings were inferior to those of another. Many societies amalgamated and became more intricately alloyed and organized; the rest, unequal to the challenge, either melted away or fled to the security of outlandish regions.

The Sorcerer. A paleolithic cave painting from the French Pyrenees.

The surviving magicomythic worlds, possessing the mightiest spirits, evolved eventually into the mythic universe of great nations, empires, and civilizations.

Those primitive societies left clinging to the archaic magic universe had no means of escape, no chance whatever against the channeled energies of the organized magicomythic universe. They vanished from the face of the Earth.

☐ ☐ ☐

Known societies, taken as a whole, display a remarkable diversity of religious beliefs, yet their various social codes of moral behavior bear striking resemblances. Nowell-Smith in "Religion and Morality" (in the *Encyclopedia of Philosophy*) comments on the religious diversity and moral uniformity of societies and draws the conclusion that moral principles are neither derivative from nor an integral part of religion.

Despite the common belief that moral codes and religious cults are inseparably related, and despite the deeply rooted notion that without religion there can be no morals, the evidence suggests that moral codes and precepts are of greater antiquity than religious cults and doctrines. The instructions given by Murray Islanders to their children, according to Alexander MacBeath in *Experiments in Living*, emphasize the importance of truthfulness, respectfulness, obedience, and kindness to parents and other relatives in deed and thought. Theft, borrowing without leave, shirking of duty, abusive language, and harm to groupmates are prohibited. Nowell-Smith remarks, "Similar lists of rules might be cited from many primitive tribes, and the lists might have come from a present-day pulpit or classroom."

Moral codes of behavior have existed since the time people began to live in social groups, and more than likely are as old as *Homo sapiens* and possibly even older. Human beings for hundreds of thousands of years and protohominids for millions of years lived in groups, and the requisite elementary codes of social behavior were thrashed out and sifted, leaving only those conducive to cohesion and mutual aid within groups. Those social groups composed predominently of untruthful, abusive, disobedient, excessively selfish, and homicidal individuals had about as much chance of surviving as the proverbial snowflake on a summer's day. Rules of collaborative and harmonious living—or moral codes—were (and still are) indispensable for the survival of a society, and primitive groups without such enforced rules were eliminated by the iron law of natural selection.

Political and religious institutions formularize the moral codes. They also invent the realpolitik excuses and sanction the exceptions to their rule. I am reminded of the story of the politician who said he was a man of principles and his first principle was flexibility. Arguments can always be found to negate moral obligations, and fortunately for society, these arguments are less durable than the primitive moral codes.

Emile Durkheim, in his monumental work *Elementary Forms of Religious Life,* rolls together all moral and religious systems and equates the lot to the constitutive structures of society. According to this grand unified theory of the social sciences, human and spirit societies were one and the same in both the magic and magicomythic universes. In pursuit of his unified theory, Durkheim exceeds what seems reasonable. Not the constitutive social structures but the human mind created the spirits. The two sides of the human-spirit mirror had equivalences and resemblances, with one side reflecting in modified form the other, yet both sides remained distinctively different, as seen by the human mind. Images observed in a mirror are not the same as the objects producing the images. As Alice found, the world on the other side of the looking glass was not the same as that left behind.

Guy Swanson in his book *Birth of the Gods* shows us that in isolated native societies there exist correlations between social structures, on the one hand, and religious beliefs, on the other, concerning ancestral spirits, mana, godlings, high gods, polytheism, and monotheism. Our simple picture of a protean magic universe evolving and branching into numerous worlds of a magicomythic universe, with each religious belief reflecting specific aspects of its society, is more in accord with Swanson's modest conclusions.

□   □   □

We have considerable difficulty comprehending the known cultures and belief-systems of primitive societies of recent times, and the problem of comprehending the belief-systems of our primitive forebears seems insuperable. The Navajo Indian does not live in our universe, and we may safely suppose that the comparatively sophisticated world of the Navajo Indian has evolved a long way from the primitive magic universe.

Within one's own universe it is extremely difficult to reconstruct a universe of long ago. The historical method recreates the past as perceived and understood within the cosmic framework of the historian. The historical facts that to us seem so tangible and important might once have been no more than episodes in the pageantry of life, whereas the intangible mind-gripping ideas that elude us were all that really mattered. Conceivably, primitive human beings had a universe that was more emotionally fulfilling and intellectually demanding than have many persons now living.

[31]

We cannot recreate the magic universe and recapture its experiences, and possibly no social group in the last 20,000 years or so has known what it was like to live in the age of magic.

Beware! Do not try to rediscover the lost heritage of entrancement. In the hall of distorting mirrors lurk images too vivid and terrifying for our matter-of-fact and orderly minds. Once trapped in the maze, the exit will not easily be found.

3

The Mythic Universe

The changeover from the magicomythic universe to the mythic universe never reached completion in North and South America, Australia, and other isolated lands secure from assault. The populations in these regions, established by the great migrations of 20,000 years or more ago, survived until recent times snug in their half-way magicomythic worlds. Elsewhere the globe was in uproar with the rise of the mythic universe.

Climate variations, animal migrations, and cultural conflicts stirred the swirl of tribal movements now swollen by population growth. In the Near East, India, China, Southeast Asia, and Mesoamerica emerged food-producing and animal-domesticating communities. Societies grew in size; institutions multiplied; economic and legal systems advanced; arts and crafts burgeoned into professions and industries; city-states and nations burst on the scene. Gigantic engineering projects, such as the construction of the pyramids in Egypt and Stonehenge in Britain, became feasible. Organized commerce, such as the trade between the Sumerian and Akkadian cities of Mesopotamia and the distant cities of Mohenjo-Daro and Harappa in India, flourished widely.

The changeover to the mythic universe was well underway more than five millennia ago, with the rise of the cosmic gods in the delta civilizations of the Nile, Euphrates-Tigris, and Indus.

"Thou are the Sole One who made all that is, the One and Only who made what existeth," chanted the Egyptian priests in adoration of Amun the god of Thebes in the New Kingdom. The mythic universe was the result of a new cosmology that ushered in a world created and ruled by almighty gods. The new gods were remote and cosmic beings: remote, for they dwelt in distant realms, and cosmic, for they sustained everything, and nothing could exist without them. But the mythic universe was much more than just the magicomythic worlds outfitted with greater gods.

At one extreme, in the old magic universe, nature throbbed with an activity of spirits; at the other extreme, in the new mythic universe, all this pulsating liveliness was withdrawn from the natural world and transferred to the gods. The natural world squeezed dry of the spirit of life became dehumanized. Human beings ceased to be entranced with the beauty and wonder of nature, and a veneration of nature changed into a veneration of gods who lived high in the sky, deep underground, or in cosmocryptic realms.

Beasts and plants and most other things in the mythic universe still displayed the same outward forms, but their nature and intrinsic meaning had changed. Trees no longer suffered pain when felled, and the need to plead for forgiveness vanished; the fire no longer was nurtured with loving care in fair return for its genial warmth and light; no need to beg permission of the wood spirit before entering the forest, the water spirit before fording the river, the bison spirit before engaging in the hunt. All was done by permission of the gods. Animals were maintained in flocks and herds to facilitate their exploitation and slaughtered in the name of the gods without apologetic ceremony.

Men and women adored and worshiped the gods of the mythic universe but not the world in which they lived; in return, the gods gathered all under their protective mantle and endowed the world with order and design. At last, by the cosmic machinations of mighty gods, it was possible for human beings to comprehend the grandeur of the created world; and by virtue of the special relations that societies and their institutions had with the gods, it was possible for human beings to influence and predict events as never before.

A worn-out magicomythic universe was traded in for a brand-new mythic universe, and though much seemed gained by the transaction,

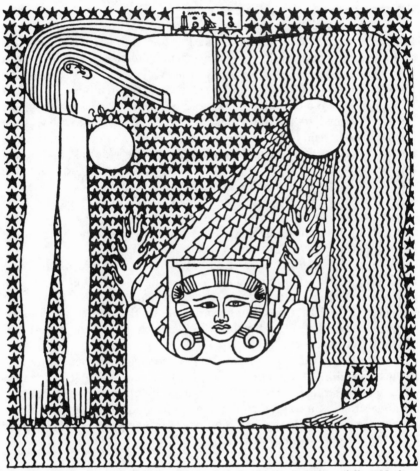

Nut the Egyptian sky goddess gives birth to the Sun, whose rays fall on Hathor the god of life and love. The Earth below is Geb, the brother of Nut.

the price paid was exorbitant. *Nature became dehumanized, and the world consisted of dead matter.* It was not assumed but known that the natural world was dead and devoid of spirit. The gods were transcendent over but not immanent within the mythic universe. True, life and spirit emanated from the gods, but such an abstract thought entertained occasionally in the temple could not be held constantly in the forefront of the mind in day-to-day affairs.

Sages for millennia have toyed with the notion that the supreme beings are immanent within the world. It is doubtful whether shoemakers and flowergirls and soldiers pursuing their ordinary interests ever took them seriously and lived accordingly. The evidence of one's senses gave proof that the world consisted of dead matter. When a person kicked a rock or cut down a tree, that person did not kick or cut down the gods; they were elsewhere and quite unconcerned. Plain for all to see was the distinction between living things and dead matter. Foremost among living things, other than the gods, came oneself, one's kinfolk, and others of one's society who worshiped the true and rightful gods. All else seemed dead, bereft of spirit, with no intrinsic spiritual value. Much too easily in the mythic universe animals were denied emotions and a capacity for feeling pain; much too easily the members of other societies worshiping different gods were denied human status and accordingly were massacred, sacrificed, or enslaved.

We see the mythic universe as a dark material world ruled above by shining gods. The utter deadness and evilness of matter stands out as its most distinguishing feature. Little wonder that in the Upanishads scriptures, Buddhist teachings, and Gnostic theologies we find an abhorrence of the dead material world, its rejection, and the advocation of world-denying asceticism.

Everything in the old magic universe behaved independently and freely. Everything in the new mythic universe behaved as if jerked into obedience by strings in the hands of heavenly puppeteers, and nothing acted naturally of its own accord: no indwelling scintillating spirit forces unless demonic, no innate self-motivating propensities unless miraculous, for all that was free was evil, and all that was virtuous was slave to the gods.

As I said earlier, the backward societies left clinging to primitive magicomythic worlds had not the ghost of a chance. They disappeared, annihilated by the organized vigor of the societies of the mythic universe and their utter disregard for human beings having other gods. The great migrations of tens of thousands of years ago into outlandish places were presumably undertaken by the tribes of the magicomythic worlds fleeing from the rising power of the societies of the mythic universe. Wherever the skirts of the mythic universe brushed against a magicomythic world, that world was obliterated or wrenched into new directions of intellectual development.

History unrolls in the age of gods as a chronicle of tyranny, warfare, human sacrifice, and slavery, disclosing the uttermost depths of human misery. This vast expanse of wretched turmoil and the loss of veneration for the natural world lie on the debit side. On the credit side lie new and grandiose concepts of cosmology, otherworldly visions of harmony and law, and abstractions that unified the universe. While gazing over the familiar historical scene, let us remember not the gods who caused untold suffering, but actually the human beings who created the gods, and by so doing organized and redirected the immense energies of the human mind.

□   □   □

Myths are about gods and goddesses, larger-than-life heroes and heroines, or ordinary persons having extraordinary experiences. The myths of the past abound with supernatural beings and supernatural events and come to us as legends in quaint and cryptic forms. Generally, they contain elements of the incredible and are viewed as entertaining superstitions or cultural dramas of sociological and historical interest. Because of the bewildering diversity of myths, we lack an all-encompassing definition of what we mean by myth. Even the authorities are at cross-purposes; social anthropologists studying Amerindian mythology do not share the views of scholars steeped in Greco-Roman classic literature, and neither are in tune with students of comparative religion.

For our purpose the simplest definition suffices. *A myth is anything from another universe that does not fit naturally into our own.* What fits naturally into the modern physical universe, as for instance, Babylonian arithmetic and Euclidean geometry, is prescient and not mythic. What fits unnaturally, as for instance, Saint Anselm's empyrean and Dante's hell in the *Divine Comedy*, is outmoded and mythic. Though incredible to us, each myth was once credible in its original setting.

Myths come to us as a legacy of priceless gems prized from their cosmic settings. Usually a myth has been recut and remounted more than once in the course of time. A full understanding of a myth requires the reconstruction of the universe in which it originated, even of the intermediate universes that modified and transmitted it, and an accurate interpretation is rarely if ever possible.

Mythology is the alchemy of myths. When societies encounter one another, intermingling their cultures, their myths react to form new mythic compounds. The conflict of gods and the victory of right over wrong depicted in myths illustrate symbolically the warfare of nations and the supremacy of one nation over another. Barbara Sproul in her *Primal Myths* describes how

> . . . these myths tell of great battles between the old, degen-
> erate gods of the conquered people and the young, energetic
> gods of the conquerors. . . . The earliest Creed of the Celts
> and the Maori Cosmologies both tell of the successful rebellion
> of divine sons against their primordial parents and reflect the
> triumph of new cultures over indigenous ones.

In Mesopotamian myths the old Sumerian female god Tiamat is defeated by Marduk, the warring deity of the victorious Babylonians, and in Hesiod's *Theogony* the male sky god of the invading Indo-European–speaking people overthrows the female earth god of the Pelasgians and Cretans.

Our knowledge of the mythic universe comes mostly from the myths recounted in annalistic literature of the ancient world—folk legends and scriptural records and oral traditions of many religions. Creation myths seem of particular interest in Western society, owing to a Judeo-Christian-Islamic inheritance that stresses the cosmic significance of these stories.

The oldest creation myths, according to Joseph Campbell in *Primitive Mythology*, draw on the generative function of the female body as their central theme. The world is a polarization of male and female elements in which the serpent and the maiden figure prominently. Neolithic cosmology makes little distinction between the organic and inorganic kingdoms, and animate and inanimate things are simultaneously created in a cosmic womb. The creative act involves all of nature, and the newborn world is a living organic whole. In the myths of later ages the living and nonliving tend to be distinguished; creation occurs as a sequential process, often as a twofold act, in which the living and nonliving are created separately, either one or the other coming first.

The 5000-year-old Sumerian epic of creation, *Enuma Elish*, tells that in the beginning, "when heaven above and earth below had not been formed," there existed the primal Apsu—a watery abyss—and the primal female being Tiamat. Apsu and Tiamat begot Anu the sky god, who with Tiamat begot Ea the earth god of wisdom. Eventually six hundred or so gods and goddesses controlled the various realms of existence, and from our matter-of-fact stance they appear to have done little more than incessantly squabble. With the rise of Babylon in the reign of Hammurabi, Ea usurped Apsu and begot with Tiamat the fearsome four-eyed Marduk. Then Marduk overcame Tiamat, divided her into the Upper and Lower Worlds, and suppressed numerous gods by appropriating their names. It was a tripartite universe of Heaven, Earth, and Netherworld, in which the wheeling stars and wandering planets unraveled the secret thoughts of gods.

Creation myths that tell of the inchoate acts of anthropomorphic and theriomorphic primal beings have much in common; all assert that without the gods there could be no universe inhabited by human beings.

According to the earliest Greek myths, as recounted in the *Theogony* (Creation of the Gods) by Hesiod of the eighth century B.C., in the beginning were four primal beings. First of all was Chaos the Limitless Void, then Gaea the Earth, Tartarus the Lower World, and Eros the Spirit of Love. In this fourfold arrangement swarms of gods were generated by the primal beings, by their matings, and by each alone, and the newborn gods personified novel theistic functions in the maturing mythic universe. A significant event was the rise of the sky god Uranus. The raping of Gaea by Uranus gave birth to the Titans, who were the first rulers on Earth, and from Uranus and Gaea came also Cronus (ancestor to Zeus), Prometheus (Forethought), and Epimetheus (Afterthought).

Out of dead clay Prometheus modeled human bodies in the likeness of gods and breathed into them the spirit of life. Hesiod in *Works and Days* tells that the earliest people were a golden race who lived free of evil and harsh toil. The earthly paradise ended when Pandora, the wife of Epimetheus, committed the original sin of inadvertently releasing the evils and diseases that Prometheus had locked away. Afterward came a silver race that neglected to worship the gods, then a strong and warlike bronze race that also perished, followed by a dishonorable and destructive iron race that still lived.

According to the Norse myths of the *Elda Edda*, out of a "yawning chasm" at the dawn of time arose the Frost Maidens, and with them came Ymir, the first of the giant gods. Their descendant, the one-eyed Odin, slew Ymir and divided his body into Earth and Sky. An apocalyptic element has entered the cosmic tales, and in the *Ragnorak* and the *Götterdämmerung* (The Twilight of the Gods) of Norse and Germanic folklore we encounter instances of the eschatological myths foretelling the end of the universe. From the beginning the world is doomed, and people and gods are destined to die in a cosmic cataclysm. The end is foreshadowed by baleful omens, oath breaking, and gargantuan warfare among gods and men. Amidst the carnage of Doomsday, the Sun becomes swollen and blood-red, and the Earth in the grip of paralyzing winter sinks back into the chasm. Out of the cosmic wreckage will one day arise a new universe, of "wondrous beauty" we are told, ruled by other and perhaps better gods.

□　□　□

One of the most impressive myths, about 5000 years old, is the *Epic of Gilgamesh* that comes from Babylonian records. Gilgamesh, a young Sumerian king of the Uruks, lives a riotous life chasing the women. In response to pleas from the citizens of Uruk, the gods create Enkidu, mortal and strong, to curb the excesses of Gilgamesh. Gilgamesh and Enkidu later become friends and share many adventures. In one episode they overcome and destroy the Bull of Heaven, and for this impious deed the gods exact retribution, and Enkidu dies. The death of Enkidu shocks Gilgamesh, and in his grief he cries out, "How can I rest, how can I be at peace? Despair is in my heart." Born of a mortal father and an immortal goddess, he himself is only half divine and therefore fated to die. Witnessing the slow corruption of Enkidu's body, he at last understands the finality of death and realizes what the denial of immortality means. He rages against his fate and its bitter injustice, "What my brother is now, that shall I be when dead," and he condemns the gods, "When the gods created mankind, Death for mankind they set aside, Life in their own hands retaining."

Far and wide he journeys seeking for the meaning of life and death, and in the course of his travels he crosses the Waters of Death to consult with Utnapishtim, the Sumerian archetype of Noah. Utnapishtim and his wife, by surviving the Flood, are the only human beings to have been granted immortality. "Because of my brother, I am afraid of death," says Gilgamesh to Utnapishtim," because of my brother, I stray through the wilderness. His fate lies heavy upon me. How can I be silent, how can I rest? He is dust and I shall die and be laid in the earth forever." But he receives no consolation, no comforting words, no meaningful answer.

The *Epic of Gilgamesh* exemplifies in a legendary figure the timeless tragedy of the death of a loved one and the enigma of one's own impending departure. Apart from the mythical scenery, it possesses today a reality as poignant as 5000 years ago.

Myths are the past reaching out to us, and we ignore them at our peril. In the *Collected Works* of Carl Jung we read, "These images are not pale shadows, but powerful and effective conditions . . . which we can only understand but never rob of their power by denying them." I am less confident than Jung about always understanding myths. Alan Watts writes in *Myth and Ritual in Christianity*, "a myth . . . is a complex of stories—some no doubt fact, and some fantasy—which, for various reasons, human beings regard as demonstrations of the inner meaning of the universe and of human life." The only reservation one might have concerning Watt's definition is that myths at the time of their inception are accounts of the credible world, and the ascription of fantasy comes later when they are half-believed or not believed.

I shall stay with my simple definition: a myth is any component taken from the world-view of another society that fails to fit naturally into our own. We may say that something *is* a myth but not it *was* a myth. The philosophical and ethical elements in myths that retain meaning, though enshrined in tales of apparent fantasy, are often real and not mythical.

We are told that our primitive forefathers were a quarrelsome lot who happily took to their clubs and spears and fought whenever possible. How else might we explain the incessant wars of recent millennia unless our genes bear the imprint of an aggressive lust? However we look at this argument, it fails to convince.

There existed little incentive for continual war among primitive social groups when the incidence of accident and disease maintained the population at a low level. Interchange of marriageable members required and ensured moderately peaceful relations. The division of activities, of armed men hunting and unarmed women and children gathering, so basic in the design of primitive society, would have been impossible if men were forever bent on raiding other social groups. Groups no doubt had their occasional differences, but ruinous and unremitting aggression seems a most unlikely characteristic of the primitive lifestyle.

The cave-dwelling Tasaday of the Philippines serve as a possible illustration of the primitive lifestyle. It is reported that they are unaggressive and their language lacks words such as enemy and war. The Tasaday seem to live in a magicomythic world not squeezed entirely dry of spirit ambience. Perhaps they are a backsliding vestige of a once vigorous society; their inoffensiveness seems to indicate, however, that aggression is not an essential trait of the primitive lifestyle.

Endemic warfare as a way of life came with the more-advanced societies and their cosmic gods. Came, in other words, with the rise of the mythic universe. Herbert Butterfield in *The Origins of History* states, "it is one of the surprises of history to learn for how long, and over how wide an area, war was a sacred thing, and was particularly associated with the action of gods." When in ancient times a monarch went to war, Butterfield states,

> . . . he would feel he was commissioned by the gods to undertake the enterprise. By appeal to the oracle or by various kinds of divinations, he would seek to know the will of the gods, taking action only at their command or when he was sure that he had their favour. It was the god who won the victory, sometimes to the discomfiture of another god.

The gods commissioned the wars, then determined their outcome by various ploys, such as depriving an army of courage or wasting it by disease.

The city-states of Sumer, each of thirty or so thousand citizens, were advanced societies operating intricate irrigation systems. Their arts and crafts had attained a high level, and by the standards of those times, the Mesopotamian culture was fabulously magnificent. Each state strived to establish the supremacy of its gods, goaded by the inspired dreams of its king and the divinations of its priests.

Wars between city-states were waged at the behest of the gods. When one state trangressed against another, by surreptitiously extending its boundaries or some other misdemeanor, it was the patron deity of the offended state who felt most involved and demanded that the citizens take to arms. Victory in battle was the reward for obedient and reverential worship, defeat the punishment for failing to observe a detail of ritual.

There were no permanent armies of trained soldiers, no carefully planned and prepared campaigns, no artfully contrived strategems; why should there be, when all was in the laps of the gods? The king who relied too much on a large army, or planned ahead too carefully, might easily lose everything as a punishment for failing to have utmost faith in his patron deity. The best insurance was to promise the god ample ceremony and sacrifice in return for victory, and in time of war, sacrifice meant human sacrifice. The fate of all wicked enemies who had the impudence to oppose and anger one's god was death, and the god could be appeased by making the wicked first suffer. In the sacred wars—every war in the mythic universe was sacred—booty and captives were the property of gods, the former went to the temple and the latter to the fire.

Astonishing as this might seem, remarks Butterfield, European history of the last 1000 years seems hardly any better, often worse, offering for our edification numerous instances of sacred wars commissioned by angry gods, of the outcome of wars determined by the will of these gods (victory because of faithfulness, defeat because of unfaithfulness), and of the severity of treatment meted out by the righteous to the monstrously wicked who opposed and angered the true and rightful gods.

Men and women have always cried out to the gods in frustration and disappointment, not because of moral injustice, but because of unrewarded service and devotion. "O God, let me see the light," implored the pious Ashurbanipal, king of the Assyrians. "How long, O God, wilt thou deal with me thus?" Well had he served the mighty god Ashur, defeating his enemies, sacrificing in all manner of ways vast numbers of captives, including young and old of both sexes, piling the dead in mountains to the glory of his Lord, and here he was in old age beset by tribulations and without just reward.

The irritable gods of the Assyrian Empire were particularly vengeful. The Assyrians were defeated in the seventh century B.C. by the renascent power of Babylon, and under the rule of Nebuchadnezzar and his successors the Persian Empire, with it more enlightened religion, rose to greatness.

The Old Testament tells of a seminomadic people racked by misfortunes and ruled by a tribal deity intolerant of all other deities. Led out of bondage in Egypt (about the twelfth century B.C.) by Yahweh, and thereafter resident on the outskirts of great empires, tossed and turned by the vicissitudes of imperial conquests, the chosen people were unified into resilient and resolute tribes by clinging tenaciously to the protection of their god Yahweh. The power of Yahweh grew in proportion not to the fortunes but the misfortunes of his tormented people. Great was Yahweh's vengeance against all who oppressed the Hebrews, and greater still against those who lapsed in their devotions. More than once in the battle songs of the Old Testament we read of Hebrew armies deliberately kept small in order that victory might manifestly be by Yahweh's decree and not the efforts of mere human beings.

The influential prophet Zoroaster (or Zarathrusta) lived in Persia in the sixth century B.C. and founded ethical monotheism. This fundamentally different version of the mythic universe supplanted the old hierarchical polytheism (akin to the Vedic religion in India) and soon became the new religion of the Medes and Persians. The moral codes of forbearance and neighborliness, which cement the elements of society, for the first time were promoted as the essential attributes of the religious life. Zoroaster was the original prophet to teach a doctrine of rewards and punishments in afterlife, in which theistic justice is tempered by moral principles. According to the teachings of Zoroaster, the Lord of Light—Ahura Mazda—created a universe in which goodness would ultimately triumph over wickedness. With the spread of Zoroastrianism came an abhorrence of human sacrifice and an adversion (less widespread) to the sacrifice of animals.

Monotheism apparently is established in either of two ways: it arises because a tribal deity is obsessively intolerant of all other deities (a situation that might occur when the constant misfortunes of a society are attributed to the demon gods of other societies), or because the topmost god in a polytheistic hierarchy is elevated out of sight, and the rest of the pantheon pales by comparison into insignificance. After the death of Zoroaster, the polytheistic hierarchy, with its angelology and demonology, staged a partial return in Persia.

During the Exile in Babylon, Jewish people encountered Zoroastrianism and adapted its apocalyptic message and ethical idealism to their own brand of monotheism. Thereafter, as revealed in the Old Testament, the duality of good and evil was stressed, and Satan, who had been an angelic minion, was assigned the role of archfiend. The Persian civilization and its Zoroastrianism inspired the Wisdom Literature of the Jews, in which good sense and principles of decent living were woven with wondrous words into the religious fabric, as found in the books of *Job* ("Where was thou when I laid the foundations of the earth? declare if thou hast understanding"), *Psalms* ("Yea, though I walk through the valley of the shadow of death, I will fear no evil: for though art with me, thy rod and thy staff they comfort me"), *Proverbs* ("Wisdom is the principal thing; therefore get wisdom; and with all thy getting get understanding"), and the *Song of Solomon* ("Who is she that looketh forth as the morning, fair as the moon, clear as the sun, and terrible as an army with banners?").

Zoroastrian idealism, with its hereditary miscellany of angels and demons, has greatly influenced Judaism, Christianity, and Islam. From the early fifth century B.C., with the fall of Babylon, until the time of Saint Augustine of Hippo in the late fourth and early fifth centuries, the Mediterranean world was exposed to Zoroastrianism through its derivative religions of Mithraism and Manichaeism and by its infiltration of Greek philosophy, Jewish prophetic literature, and Gnostic and Neoplatonic theologies. Augustine, who molded Western Catholicism, was at first a Manichaean, and after his conversion he blended Zoroastrian ideals with Judaic scriptural history. In *The Eternal City* Augustine compared the Heavenly and Earthly Cities and contrasted otherworldiness and the way of grace and salvation with worldiness and the way of evil and damnation.

Gods whose cults included ethically inspiring elements, such as Osiris and Isis (the divine mother holding her child), were not uncommon in the ancient world. But the novel concept of a supreme godhead as absolute goodness, full of compassion and concern for human beings, originated in the pastoral milieu of Persia, and its highest ideals are exemplified in the scriptural accounts of the life of Jesus of Nazareth.

With ethical monotheism came the paradox of a beneficent supreme being responsible for creating a universe containing evil. As Alan Watts remarks in *The Two Hands of God*, "This, then, is the paradox that the greater the ethical idealism, the darker the shadow we cast, and that ethical monotheism became, in attitude if not in theory, the world's most startling dualism." The more we exalt and glorify the gods, the darker becomes the material world by comparison, and the worse we seem ourselves.

☐   ☐   ☐

"The conception of gods as superhuman beings endowed with the powers to which man possesses nothing comparable in degree and hardly in kind had been slowly evolved in the course of history," wrote James

Frazer in *The Golden Bough*. Frazer discussed the evolution of animism into theism and speculated on how the management of the "gigantic machinery of nature" was handed over to the gods. He suggested that the transition to the god-created and god-controlled mythic universe was partly the result of a conscious decision made by those having "acuter minds." This is almost as unlikely as Edward Tylor's earlier suggestion that animism was the invention of "savage philosophers." Once the spirits had emerged, then automatically in the course of time, as knowledge broadened and deepened, the spirits acquired the trappings of power, and the more that societies depended on them for guidance and protection, the greater in proportion grew their power.

Frazer traced the roots of religion to the transition from animism to theism and suggested that religion did not exist before the advent of gods. This point of view depends on what is meant by religion. It seems safe to say that the emotional experience essential to a religious way of life was rife in the age of magic, and an unbiased person might be pardoned for thinking that religion in some form or other is as old as *Homo sapiens*.

Alfred Whitehead, philosopher and mathematician, in *Science and the Modern World*, has this to say on religion: "It is the vision of something that stands beyond, behind, and within ... yet eludes apprehension; something whose possession is the final goal, and yet is beyond all reach; something which is the ultimate ideal, and the hopeless quest." These appealing words, referring to the unattainable, strike a responsive chord but unfortunately fail to help us much. With equal effect they apply to the goals of art, philosophy, and science.

Even the proverbial intelligent schoolchild knows that religion is difficult if not impossible to define. In a myriad guises religion distinguishes between the sacred and profane and aspires to unite both in a cosmic scheme. The magic and mythic universes stand at the extremes of a religious spectrum; at one end, in the magic universe, nothing or almost nothing in the surrounding world is profane; at the other, in the mythic universe, everything or almost everything in the tangible Earthly City is profane, and everything in an intangible Heavenly City is sacred.

Various schemes of classification have been suggested. One particular scheme, of universal scope, classifies all religions into three main groups: the prophetic religions (such as Confucianism, Mohammedanism, Judaism, Protestantism), the sacramental religions (such as Catholicism, Hinduism), and the mystical religions (such as Buddhism, Sufism, Quakerism). The prophetic group stresses the value of belief and doctrine; the sacramental group, the value of ritual and doctrine; the mystical group, the value of emotional experience and the unimportance of doctrine. Many persons will regard this as a simplistic scheme, for most religions have prophetic, sacramental, and mystical components, including ethical ingredients, and function in accordance with what their devotees perceive as the proper distinctions between the sacred and profane.

I shall sweep aside all descriptive classifications and take the sophomoric view that the basic elements of religion, of comparable value, are ideas and feelings. Thus religion is both conceptual and emotional: conceptual—consisting of ideas—in a collective or social sense and emotional—consisting of feelings—in a personal sense. On the one hand lie religious beliefs and doctrines, usually shared and inherited, that contribute to the wealth of social phenomena; on the other hand lie religious feelings and emotions, peculiar to individuals, that contribute to the richness of personal life. The ideas evoke mystical feelings and canalize religious emotions. In art that aims to expose the exquisite with the highest skills, a superstructure of abstract ideas is not vitally essential. In philosophy that aims to understand by means of critical discourse, and in science that aims to understand by awakening the dead world of matter with activating harmonies, a substructure of feelings is not vitally essential. Religion is thus unique in that it makes demands on the whole person, intellectual and emotional, social and personal, and consists of an interplay of thoughts and emotions of equal significance.

My sophomoric viewpoint leads to the following conclusion. Religious ideas, which depend on the structure of universes, change from society to society, but religious feelings, whose intensity depends on the human capacity for emotional experience, possess much in common throughout all societies. Concepts relating to objects of veneration and modes of worship change from society to society, but the associated emotions have much the same intensity in all societies. Changing ideas and unchanging emotions are both vitally important in religion. Adherents of "true" religions will, of course, disagree, for they hold that their fixed ideas are imperishable and god-given.

The conflict between science and religion focuses always on religious ideas and never on religious emotions. When devotees of various religions insist on retaining their cosmic belief-systems, contrary to the evidence of science, they commit the error of supposing that religion consists of imperishable myths and of believing that without mythology they cannot have religion. Rejection by science of mythology brings science into conflict with religious ideas but not religious feelings.

If religion is to retain its twofold character and remain distinct from art, philosophy, and science, while fulfilling its role of unifying society and sustaining the individual, then its ideas and beliefs must forever change in such a way that contemporary modes of thought remain evocative of religious emotions. When denied this twofold character, religion withers, and the universe loses personal meaning. Phony cults, upstart mysticisms, and professional opportunists usurp its role with sorry consequences. The "rights of man" ensnared in a legislated web of regulations, for example, are championed with religious fervor, and serve as substitutes for old social codes of honor, charity, and self-imposed moral discipline; society becomes a burden of laws (what is not compulsory is forbidden), and in protest the individual turns lawless. Religion has no substitute; it is as old as society and hence as old as *Homo sapiens.*

□   □   □

Religion in a rudimentary and spontaneous form existed in the animated magic universe. On passing from the magicomythic worlds into the mythic universe we see ideas of increasing abstraction becoming evocative of religious emotions. Ideas on divinity but not things of nature are then worshiped and venerated. Abstract ideas become self-sufficient, resting on a foundation of faith and not everyday emotional experiences, the sort of faith that accepts the validity of myths without supporting evidence. Religion drifts away from its primitive emotional basis, tending toward a onefold instead of a twofold state, and becomes no more than theoretical veneration of a social institution.

I have in mind Cardinal Nicholas of Cusa in the fifteenth century, sitting in his cloistered study, meditating deeply on the omnipotence of the supreme being, and having ideas about the universe that anticipated modern thoughts. At the same time this intelligent man organized from his study the torment of Jews and the persecution of heretics. All persons failing to conform to his own institutional religion, however sincere in their beliefs, ranked as subhuman beings. Ideas, not emotions, were all that mattered.

What are the gods? We can do little better than take the view of the Ionians, who contemplated them with disinterest and sought by inquiry to discover what truly deserves veneration. No cosmologist knows what is the Universe, and no theologian knows what is God. I discussed the difference between Universe and universes in the first chapter *(Introducing the Masks)*, and I shall discuss the parallel difference between God and gods in a later chapter *(The Cloud of Unknowing)*. It suffices here to say that the word god is used to denote a model of God, in the same way that the word universe is used to denote a model of the Universe. The many universes serve as the masks of the Universe, and the many gods serve as the masks of God. I intend no disrespect when some cherished theistic model, such as the Christian supreme being, is referred to as a god. We must allow for the gods of other religious beliefs and the possibility that better models may yet come.

I am aware of the questions this raises, and I attempt to answer a few in *The Cloud of Unknowing*, but not in a way that will please everybody in Western society. In short, I follow the Spinozistic solution of equating God and Universe, thus radically departing from the historiography of the mythic universe, and in this way find that I must adopt the mystical view that the Universe and God are everything, forming a UniGod beyond human understanding. Thus the dead mythic universe is overthrown and the world reanimated with wondrous magic.

□   □   □

A myth is a remnant torn from another universe that cannot be patched naturally into our own. That is roughly what I have said, and I must now stand back and look at the subject more generally. Myths are extremely important. I doubt whether our society and hence ourselves could exist without them.

The universe of our society is the modern physical universe. Other periods of history have had their universes, their intellectual vistas and wide avenues of creative thought. In this century the physical universe embodies our most advanced and effective ideas and is the main outlet for disciplined and creative thought. The physical universe is the world that best characterizes our period of history. It soars around us, revealing the inner structure of cells and stars, dominating our lives, directing the destiny of nations. The gods of the mythic universe are myths because they fail to fit naturally into the modern physical universe.

But the physical universe is the *Weltanschauung* of our society and not the world in which individuals live every moment of the day. It is neither my world picture nor yours by a long shot. Human beings cannot live fully immersed in the abstract universe of their society. The world pictures in which they live are a ragbag of beliefs taken from the past and patched together in the present. Patchwork world pictures are much more heterogeneous and complex than universes. This is just as well for two reasons.

First, despite its coherence and effectiveness, the physical universe is much too simple in which to live; it is incapable of guiding us in our personal affairs and the problems of running society. We fall back on the mythic elements in our world pictures for codes of behavior, notions of right and wrong, concepts of justice, freedom, duty, love, beauty, and religion, all of which lack meaning in the physical universe. We render unto the physical universe what is physical and unto the mythic universe what is mythical. What is mythical in the physical universe may not be meaningless in the world pictures of individuals.

Second, the physical universe is incomplete, and it may be fundamentally wrong as judged by the universes of the future. It is itself doomed to become mythical. We live in our quasi-mythical world pictures that plumb the depths of the mind, but the mind is only partially revealed in the lucidity and effectiveness of the physical universe. World pictures are wellsprings from which gush the ideas that continually transform the universe of our society. If we lived entirely immersed in the societal universe, the wellsprings of inspiration would dry up, and like termites we would be intellectually dead.

When wearing our scientific hardhats we deplore the credulities of the mythic universe. But the magicomythic and mythic universes evolved in step with the development of society, and if we deny the importance of their beliefs, society must collapse or fall prey to the mad universes that destroy world pictures. A philosophy that uproots the past and denies the validity of all mythic principles is hell-bent for state slavery and a termite-like society; whereas a philosophy that sees people as religious and moral creatures, having duties to one another and freedom to believe and practice the precepts of their world pictures, is more in accord with our primitive origins and the cohesive principles of a society of intelligent beings.

☐ ☐ ☐

The age of gods withdrew the spirits of the magic universe and gave us in return a vision of unity and harmony—gave us also the dichotomy of living beings and dead matter in a valley of shadows pierced by pale theistic rays.

Perhaps the mythic universe was a mistake. Perhaps there was more than one way for human beings to evolve, and 10,000 years or so ago by misadventure we took the wrong turning, following a path that dehumanized the world, that may in the end lead to a globe reduced to rubble. By taking a different path we might now be out there among the stars.

4

The Geometric Universe

Four thousand years ago the Babylonians charted the heavens, divided the sky into the constellations of the zodiac, compiled star catalogs, recorded the movement of planets, prepared calendars, and predicted eclipses of the Moon. Though skilled in the arts of computational divination, the Babylonians did not theorize on the laws of celestial harmony, for they were not scientists but priests paying homage to the mythic universe.

About the sixth century B.C. intellectual activity quickened in many lands. The teachings of Zoroaster in Persia, Gautama the Buddha and Mahavira the Jain in India, and Confucius and Lao-tzu in China gave birth to ethical doctrines and inspired religions of virtuous living. Meanwhile in the Hellenic world a movement of a different kind had begun that also would lead to eventful consequences.

The Greek civilization of scattered cities and colonies was a mosaic of cultures that nurtured an unusual elasticity of mind. Greek philosopher-scientists of the sixth century B.C. developed a style of thought radically different from the mystery-mongering of Babylonian and Egyptian astrologer-priests. The Greeks dissected the world and speculated on the nature of its elements and their interactions; they awoke the dead matter of the mythic universe, thus enabling them to disentangle sequences of cause and effect in a world of natural happenings; they invented the rudiments of the scientific method, looked askance at the sacred myths, and to this day science inherits their curiosity and incredulity.

It began with the Ionians, descendents of the Mycenaeans, who had inherited the culture of Minoans of Crete. In "the midst of the wine-dark sea," in the words of Homer, 1000 years before the Ionians, the Minoan civilization reached the pinnacle of its power and glory. It was Europe's first civilization. Little is known of the Minoans and nothing of their origin; their language was not Indo-European, and their Linear-A script remains undeciphered. The Minoans were an unwarlike maritime people, living in unfortified palaces having spacious courts, bathrooms, toilets, and drains, but no large temples, and their colorful frescoes of animals, birds, and fish display a spontaneity at variance with the stylized art of Egypt and Mesopotamia. This lively and supple culture—Hesiod's silver race that neglected to worship the gods?—expired rather suddenly, perhaps because of a violent volcanic eruption on the island of Thera, and also because of the incursions of warlike Mycenaeans. The Mycenaeans, rich in gold earned as mercenaries while aiding Egypt in the ejection of the Hyksos, absorbed the surviving Minoans and their culture; they later defeated Troy, as related in the Homeric epic, then in the eleventh century B.C. withdrew from the mainland of Greece to evade the ravages of invading Dorians.

Whatever spiced Minoan life may have descended to the Ionians of the sixth century B.C. and inspired their acuter minds into revolt against the mythic universe.

Thales, born in the late seventh century B.C. and first of the Ionian philosopher-scientists, lived in Miletus on the Turkish Mediterranean coast. His interests ranged widely, and he had an original style of thought; whenever possible he used geometry learned from the Egyptians, and he predicted an eclipse of the Sun with astronomy learned from the Babylonians. The world floats in a primordial sea, said Thales, and is itself composed of water manifesting in many forms; water is the primary element and the ultimate constituent of all things, for it lives, flows with independent movement, and permeates and animates the world.

All is "according to necessity . . . and the assessment of time," said Anaximander of Miletus, a disciple of Thales. He argued that no single substance may be regarded as primary, for the ultimate is unbounded and its nature indefinable. He proposed the idea that the world consists of intermingled opposites—hot and cold, dry and wet, light and dark—and is animated by their interplay. He was, observed Agathemerus, the first "who dared to draw the inhabited world on a tablet," in other words, to make a map of the known world. Anaximander taught that the world alternates over long periods of time between extreme states and that animals, including human beings, have evolved from primitive creatures in the sea.

We know, said Anaximenes who lived also in Miletus and was a pupil of Anaximander, that air is pervasive and forever restless. Air is the breath of life and must therefore be the ultimate substance. Air is flame and fire when rarified, cloud and water when condensed, earth and rock when more condensed. Condensed air is cold, rarified air is hot, and we notice, he added, how our breath feels cool when forced between pressed lips, yet is warm with the mouth open. Anaximenes said the constellations of stars are fiery rarifactions high above the atmosphere.

Unlike the Egyptians and Babylonians, the Greeks in the sixth century B.C. lacked reliable and continuous historical records. The legendary past when gods walked the Earth was separated from them by an inpenetrable dark age. Hecataeus, born in Miletus while Anaximander and Anaximenes still lived, traveled widely and became the founder of geography and the first critical historian. "The tales told by the Greeks are many and in my view ridiculous," he wrote. As a young man, Hecataeus told the priests of Egypt that the Greeks could trace their ancestry back for as many as ten generations, and even more, to a time on Earth when human beings were still godlike. The Egyptians laughed, and showed him the statues of their high priests, one after another, arrayed in serried ranks, extending back for thousands of years. The astounded Hecataeus thereupon began a career of systematic investigation that established history as a disciplined study.

According to Plato in the *Timaeus*, when Solon, a poet and statesman of Athens, visited Egypt in quest of the past, he was told by an old priest, "Solon, Solon! You Hellenes are perpetual children. Such a thing as an old Hellene does not exist." Then, referring possibly to the Minoans, the priest said:

> You have preserved only the memory of one deluge out of a long previous series. . . . You are ignorant of the fact that your own country was the home of the noblest and the highest race by which the genus *Homo* has ever been represented. You yourself and your whole nation can claim this race as your ancestors through a fraction of the stock that survived a former catastrophe, but you are ignorant of this because for many successive generations the survivors lived and died illiterate.

Herodotus, the "father of history" had a similar experience, and after Hecataeus, every Greek student of history spent a semester in Egypt.

The Ionians initiated Greek prose writing and raised the Hellenic arts to a high level. With unfettered curiosity they peered into the structure of matter, pondered on the nature of time, conjectured on the distinction between planets and stars, studied geological and biological evolution, developed metreology, and theorized on the mechanics of storms. The world-shaking ideas of the Ionian thinkers no doubt caused amazement among their less-distinguished compatriots, who had at least the merit of forbearance. Our knowledge of the Ionians is slim and fragmentary; how they acquired a disinterested attitude toward the gods, and why they were not burdened with the usual politically powerful priesthood, we do not know.

Heraclitus of Ephesus lived in the late sixth century B.C. and was an Ionian of a different stamp. Like most Greek thinkers at this time he was influenced by Pythagoras, whom we shall come to shortly. Heraclitus taught that the Logos—the Word or God—was the unifying principle. He is best known, however, for advocating fire as the primary element (the world "was ever, is now, and ever shall be an everliving fire"), and for his system of perpetual flux ("all things change and nothing remains at rest," and we "cannot step into the same river twice"). Reality is process, being is forever becoming, and wisdom consists of nothing more than knowing how things change; only change is changeless.

Heraclitus conceived a universe of flux animated by a conflict of cosmic forces, in which "harmony consists of opposite tensions like the bow and the lyre"—a world at war with itself, having neither peace nor rest, torn by strife, and sustained by the enmity of its elements. "We must know that war is common to all, strife is justice, and all things come into being and pass away through strife." With moderate certainty we may venture that Heraclitus anticipated the notion of "survival of the fittest." The Heraclitean system of whirling bodies and swirling fluids never at rest foreshadowed aspects of the Cartesian mechanistic universe.

□   □   □

From the Ionians the scene changes to the Eleatics in the Greek city of Elea on the southern coast of Italy. Parmenides, most prominent among the Eleatics, had little sympathy for the Heraclitean system of perpetual flux. Nothing changes, he said; all change is mere appearance, and wisdom consists of rejecting the world of transient happenings. Reality and being are timeless, and process and becoming are illusions of the deceived senses. Beyond the phenomenal world of dissonant tumult lies an "invariant sphere of being," an immaculate reality that can be reached and grasped by reason alone. Parmenides was the first to make the duality of appearance and reality the basic issue. What we experience—a welter of sights and sounds—is appearance only; beyond and behind, hidden from human view, lies ultimate truth. Since the time of Parmenides the fundamentals have been timeless. Absolute truth, beauty, and goodness never change.

We know little of Parmenides and his philosophy. No doubt he said much more and tried to explain why his universe—the invariant sphere of being—was unchanging. With hindsight we understand that it was timeless for the simple reason that it contained time (see *The Fabric of Space and Time*).

The Parmenidean universe anticipated the modern physical universe of spacetime. Consider the following. Events at places in space occur at moments in time; they are displayed in space and time and therefore are fixed and unalterable. The place of one's birth is fixed in space and cannot change; the date of one's birth is fixed in time and also cannot change. Events of the past, present, and future are displayed in time and cannot change; only our movement through time makes them appear to come and go, to change from moment to moment; permanent being assumes an appearance of becoming, fixed reality assumes an appearance of process. As individuals we perceive at any moment only a fragment of the whole universe, but as we move through time, the universe unfolds, revealing more and more of what timelessly exists and previously was unknown. Suppose we were not chained to our sense and could see the whole universe displayed throughout space and time: there would be no unfolding, no state of change, no appearance of becoming, for all would then be timelessly disclosed in a single all-embracing observation.

This argument, advanced in the nineteenth century before the theory of relativity, explains the spacetime continuum from the perspective of an experiencing individual. But how we move through time and how we experience a sense of change and an awareness of becoming in an invariant sphere of being—or spacetime continuum—is still not understood. Some scientists and philosophers regard our awareness of change as an illusion, of psychic significance only, and like Parmenides believe that becoming is not fundamental in the physical world.

The systems proposed by Heraclitus and Parmenides represented the extremes of action and inaction. The frenetic strife of Heraclitean flux, on the one hand, and the deathlike stillness of the Parmenidean sphere of being, on the other, demanded some attempt be made to reach an acceptable compromise. Empedocles of Acragas in Sicily, to this end (but not very successfully), introduced love as a moderating force. Love attracts and unites, strife repels and divides, said Empedocles, and the everlasting elements of earth, water, air, and fire are governed by the sway of love and strife. "Nay, there are these things alone, and running through one another they become now this and now that, and yet remain ever as they are," and from the convoluted forces of love and strife flow "forth the myriad species of mortal things, patterned in every sort of form, a wonder to behold."

Anaxagoras, born in Clazomenea near Ephesus about 500 B.C. just before the cities of Asia Minor fell under the rule of the Persian Empire, was among the last of the outstanding Ionian philosopher-scientists. He lived and taught in Athens at the time of Pericles. "Nothing comes into being or perishes, but is compounded from or dissolved into things that are," said the renowned Anaxagoras, and the "things that are" exist everywhere in minute portions. More than likely he inspired the atomists by arguing that all things are compounded from numberless minute portions of an elemental and universal substance.

The Moon, said Anaxagoras, shines by reflected light and has mountains on its surface; the stars are fiery bodies so distant that we cannot feel their warmth. He originated and launched the momentous idea of a universe of unlimited extension, in which things everywhere are similarly composed and subject to similar laws. How such world-shaking ideas originate in the human mind defies understanding. The universe is ruled by Mind—the Logos—not the surd gods, said Anaxagoras, and for this heresy against the mythic universe he was impeached (Athenians were less tolerant than Ionians), and though acquitted, owing to influential friends, he deemed it wise to flee the hostility of Athens.

The atomist universe in the fifth century B.C. was conceived jointly by Leucippus (of whom nothing is known and who may have been among the last of the Ionian thinkers) and his follower Democritus, who taught at Abdera in Thrace. This novel universe consisted of countless atoms (particles that supposedly could not be further subdivided) distributed in a void of infinite extension. Atoms are made of a universal primary substance, said the atomists, and differ only in shape and size. Our sense impressions of color, sound, smell, touch, and taste reside not in the things themselves, but are qualities of a secondary nature originating in the sense organs. "By convention there is color, by convention sweetness, by convention bitterness, but in reality there are atoms and the void," announced Democritus. The universe consists only of atoms and the void; all else is opinion and illusion. If the soul exists, it also consists of atoms.

Endlessly and freely the atoms moved through the void, repeatedly colliding, occasionally sticking together, aggregating into bodies of various configurations that evolved and finally dissolved back into the atomic ferment. Strewn throughout the atomist universe were numberless worlds with orbiting moons; on each world life originated as primitive organisms and evolved to a civilized state. To this day the sweep of the atomists' vision startles and holds the imagination.

In his poem *The Nature of the Universe*, composed in praise of the atomists and Epicurus, the Roman poet Lucretius in the first century B.C. wrote:

> But multitudinous atoms, swept along in their multitudinous courses through infinite time by mutual clashes and their own weight, have come together in every possible way and realized everything that could be formed by their combinations. So it comes about that a voyage of immense duration, in which they have experienced every variety of movement and conjunction, has at length brought together those whose sudden encounter normally forms the starting-point of substantial fabrics—earth and sea and sky and the races of living creatures.

Echoing the atomists, Lucretius declared, "Bear this well in mind, and you will immediately perceive that nature is free and uncontrolled by proud masters and runs the universe by herself without the aid of gods." In response, the gods of the Greco-Roman world became, if anything, more powerful and inscrutable than before.

☐   ☐   ☐

In the background, hitherto ignored, stands the enigmatic figure of Pythagoras. Born on the Ionian island of Samos in the early decades of the sixth century B.C., and perhaps a student of either Thales or Anaximander, Pythagoras is reputed to have traveled widely and imbibed knowledge from many sources. In his later years he taught at Croton in the south of Italy and founded a society similar to the Orphic communities then flourishing in Italy and Sicily.

According to Diogenes Laertius, Pythagoras was "the first to call the heavens cosmos and the Earth a sphere," the first to call the universe a harmonious whole and the Earth a spherical body. The universe resembles a finely tuned musical instrument, said Pythagoras, and the celestial spheres, governed by geometric laws, move with musical harmony in circular paths about a central, unseen cosmic fire.

Pythagoras, more than anybody else, founded mathematics. He formulated theorems with economy and rigor and developed geometry to the level at which Euclid inherited it. By experimenting with vibrating strings Pythagoras found the arithmetical relations of harmonious notes,

confirming his conviction that beneath the noisy tumult of common oc-
currences dwelt a harmony of numbers. The Pythagoreans—followers
of Pythagoras—believed that by decomposing the world into numerical
quantities they revealed its true nature and innerrelations. Like modern
theoretical physicists, they paid homage to a universe suffused with
arithmetic divinity.

The word *philosophy*, meaning "love of wisdom," comes to us from
the Pythagoreans, who sought wisdom with passionate enthusiasm.
Whereas the Ionians in their disinterested pursuit of knowledge had
cold-shouldered enthusiasm (a Greek word meaning "possessed by the
gods"), most of the Pythagoreans had little fear of prudence being swept
aside by enthusiasm. Bertrand Russell, mathematician and philosopher,
comments:

> To those who have reluctantly learnt a little mathematics in
> school this may seem strange; but to those who have expe-
> rienced the intoxicating delight of sudden understanding that
> mathematics gives, from time to time, to those who love it,
> the Pythagorean view will seem completely natural even if
> untrue. It might seem that the empirical philosopher is the
> slave of his material, but that the pure mathematician, like
> the musician, is a free creator of his world of ordered beauty.

Apparently, the Pythagoreans were influenced by Orphic beliefs, derived
from the myth of Dionysus, according to which the immortal soul is
divine and revelation the source of religious experience. The Orphic creed
in this form bore little resemblance to the worship of Bacchus, and possibly was yet another echo of the Minoan culture. *Orgy* is an Orphic
word signifying "sacrament that purifies the soul," whose meaning later
became debased by association with Bacchic revels. *Theory* and *theater*
have the same root, meaning "to view." *Theory* came into philosophy
and science through the Pythagoreans and was used in the Orphic sense
of "passionate sympathetic contemplation."

Scientists supposedly are dispassionate in accordance with Ionian tradition, yet a cold-blooded attitude is rarely the entire truth. Scientists
generally pursue their theories with passionate dedication, possessed
by enthusiasm, as in Orphic ecstasies and orgies, and with enthusiasm
deny their enthusiasm. Theorists usually are otherworldly, like the Pythagoreans, contemplating passionately an abstract reality unified by
numbers, and viewing the phenomenal world as ephemeral and at best
suggestive of concealed immaculate harmony. Albert Einstein, a twen-
tiety-century Pythagorean, said:

> The most beautiful emotion we can experience is the mystical.
> It is the sower of all true art and science. He to whom this
> emotion is a stranger, who can no longer wonder and stand

in rapt awe, is as good as dead. To know that what is impenetrable to us really exists, manifesting itself as the highest wisdom and the most radiant beauty, which our dull faculties can comprehend only in their most primitive forms—this knowledge, this feeling, is at the center of true religiousness. In this sense, and in this sense only, I belong to the ranks of devoutly religious men.

☐ ☐ ☐

Protagoras, a Thracian born in Abdera and a contemporary of Socrates in the second half of the fifth century B.C., was foremost of the Sophists who made their reputation by teaching philosophy and expounding the art of rhetoric. Skill in rhetoric and an ability to prove black is white were and still are indispensable to all seeking legal advantage and political advancement. They have other uses, as Samuel Butler the satirist said of the clergymen who switched their professed beliefs while Charles I, Cromwell, and Charles II moved in and out of office:

> What makes all doctrines plain and clear?—
> About two hundred pounds a year.
> And that which was prov'd true before
> Prove false again? Two hundred more.

Two hundred pounds in the seventeenth century was the annual income of a comfortable clerical living. Protagoras would have approved of this game of musical chairs. He thought the gods probably did not exist, but a prudent person should hedge his bets by worshiping one or more—it did not matter which. "Man is the measure of all things," he said, which contains much truth and yet is a dangerous weapon in the wrong hands.

The Sophists had little patience with Pythagorean mysteries, and their philosophy, lacking firm conviction in the truth of ultimate reality, led directly to skepticism. What is good in one society is bad in another, argued the Sophists, and hence ethical principles are relative and without absolute value. What is right and proper for one person is wrong and improper for another, and nothing is right or wrong but thinking makes it so. Arguing thus, with their theory of cultural and ethical relativity, they fostered the view that gratification of one's desires is all that matters, similar to the view prevailing today.

Socrates, who lived in Athens and was sentenced to death in 399 B.C. on a trumped-up charge, devoted his abundant energy to countering the Sophist doctrines. Self-knowledge is wisdom and the doorway to serenity of mind, he taught, and by inner searching we discover the distinctions between right and wrong. "Virtue is knowledge" gained by introspection, said Socrates, and evil is ignorance of the knowledge that lies latent within us. He believed that truth dwells within ourselves awaiting recognition. The Socratic method of inquiry consisted mainly of asking questions, one after another, and the interrogated person discovered, step by step, knowledge already possessed yet previously unrecognized.

In Plato's dialogue *Phaedo* we find Socrates much dissatisfied with Anaxagoras, because, when Socrates was a young man, the Ionian did little more than explain *how* the universe works. Socrates thought he should have shown more interest in *why* things exist and are for the best. Science and philosophy, as distinct subjects, began with the Greeks, and it is therefore interesting to see how Socrates distinguished between the two and the reason why he favored philosophy. His argument, freely interpreted and much adorned, is as follows.

Science deals with questions of the how-type, whereas philosophy tries to dig deeper and deals with questions of the why-type.

How does a bird fly? How does a fish swim? Both are scientific questions. Discoveries in science are often the result of asking the right question at the right moment, and the relevant question with its novel viewpoint is usually of the form, How . . .? Superficially, How . . .? and Why . . .? have little difference and are often interchangeable. Thus, Why is the sky blue? may be changed to, How does the atmosphere scatter sunlight? The latter is more scientific than the former, however, for it stresses the functional rather than the existential. Why is a thing the way it is? and How does it function? are typical scientific questions, but the first may be rephrased more to the point by asking, How did its components come together and evolve into its present particular state?

Sometimes a why-type question cannot be rephrased without considerable alteration in meaning, and then it is doubtful whether the question makes much scientific sense. Thus, Why do atoms exist? is not equivalent to How do atoms work? The first is philosophical and the second scientific. Fundamentally, the how-type question (how things assemble, how they work, and how they evolve) is functional and usually scientific, and the why-type question (why things exist and why they are best) is existential, or ontological, and usually philosophical.

Scientists advance postulates, and if the consequences deduced from the postulates are in accord with observations, that is their sufficient reason. Socrates explained to Cebes why he disliked this attitude; as a philosopher he felt interested mainly in why-type questions and wanted to know more than mere functional descriptions of how things work.

The Ionians usually did not perplex themselves with the problem of why the postulated elements exist; instead, they asked how the postulated elements function, how they change their form, and how they account for what is observed. Socrates, Plato, and the philosophers who followed in their footsteps were far more concerned with the reason and necessity of things and with why reality is the way it is. The Ionians and the scientists who followed in their footsteps attended to the more prosaic task of explaining how things function, taking the view that we never know why reality is necessarily the way it is and can only know how things actually work; all else is for philosophy and theology. The early scientists used mechanical and functional analogies drawn from the arts and crafts, and as the poet Berton Brayley said, "Back of the beating hammer . . . the seeker may find the thought."

I must interject the following remark. All universes in the past were seriously in error, as we know, and they seem therefore to hang in the air, unsupported with facts obtained by observations. We think only our own universe is inductively secure, and the rest were no more than deductive fancies. We feel tempted to think that the Greeks reasoned deductively from general principles assumed to be self-evident, never inductively from observations, that they were theorists like Socrates, never experimenters and observers. This temptation must be resisted. From Thales studying the nature of water, Pythagoras investigating the resonances of vibrating strings, Hippocrates establishing the methodology of medicine, Aristotle dissecting the external world, Archimedes inventing levers and mechanical contrivances, Eratosthenes measuring the diameter of the Earth, and hundreds of other examples, to the pumps and steam engines and research projects at the Museum in Alexandria, a history of empirical investigation unfolds without which there could have been no science.

After the Ionians science and philosophy became divergent streams of thought. The scientific stream, empirical and realist, passed through Aristotle; the philosophical stream, rational and idealist, passed through Plato. They merged in the medieval universe and then separated again in the mechanistic universe.

☐　☐　☐

The two-sphere universe was popular among Athenian scholars in the early fourth century B.C. at the time of Plato. It consisted of little more than a spherical Earth surrounded by a distant celestial surface studded with the stars of heaven. The Earth, the inner sphere, stood motionless at the center of the universe; the heavens, the outer sphere, rotated daily. The two-sphere model was a makeshift Pythagorean universe, stripped down and modified to accommodate explicitly the anthropocentrism of the mythic universe.

The heavens rotated about a central and stationary Earth. Overhead, beneath the stars, the planets wandered in strange and peculiar ways. At the instigation of Plato at the Academy in Athens, this simple picture was elaborated in an attempt to explain the planetary paths. To support the planets, additional spheres were introduced, and these new and intermediate spheres rotated at various rates about different inclined axes. In this way the two-sphere model became a many-sphere system, which at first was of little more than academic interest.

Aristotle, of Stagira in northern Greece, studied at the Academy until the death of Plato; he then traveled and at the Macedonian court was appointed tutor to Alexander, a youthful firebrand who later became king of Macedonia. While Alexander was conquering the Middle East, the Persian Empire, and places more remote, Aristotle returned to Athens and founded his own school known as the Lyceum. Aristotles's lectures ranged widely, covering natural history, biology, physics, logic, politics, and ethics, filling 150 volumes.

Aristotle took the many-sphere model and invested it with physical and spiritual reality. The scribblings of the geometers blossomed into a full-blown universe. The planets, including the Sun and Moon, starting from the innermost, were the Moon, Mercury, Venus, Sun, Mars, Jupiter, and Saturn, and each had a supporting system of linked crystalline spheres. Altogether fifty-four spheres were needed to make it work. It was a geometrical universe, geocentric, and of finite size, extending to the outermost sphere of stars. The ceaseless rotations of the many spheres, said Aristotle, have persisted throughout eternity.

In the Aristotelian universe the physical elements of earth, water, air, and fire were the tangible constituents of the Earth and the sublunar region. The heavenly bodies and their supporting spheres consisted of a fifth element, called ether, which was intangible and nonphysical. The natural motion of the physical elements was upward and downward, as they sought to find their place according to weight, and all other motions tended to be violent in character. The natural motion of the etheric element was endless perfect rotation around the Earth, which was poised at the center of the universe. It is fitting, commented Aristotle, that the physical elements of perishable forms should have imperfect and incomplete motion, away from and toward the center of the universe, and this explains why the Earth does not rotate. It is equally fitting that the etheric element of imperishable forms should have perfect circular motion, and this explains why the heavens forever rotate around the center of the universe.

Generation and decay occurred only in the physical realm of the Earth and its immediate sublunar regions. Above the sublunar realm everything remained changeless and perfect.

Comets and whatever else marred the perfection of the heavens

Aristotelian system, with the Earth at the center of the universe, surrounded by the celestial spheres. The primum mobile, outside the sphere of fixed stars, is an Arab addition.

were omens occurring within the sublunar realm and little more than atmospheric phenomena. This notion persisted for two thousand years, and whenever a new star flared in the sky, astonished observers shook their heads in disbelief. The supernova of A.D. 1054 in the constellation of Taurus, which was bright enough to be seen in daylight according to Chinese chronicles, was not recorded by Western Europeans; they could not believe what they saw; it was unorthodox, and therefore not worth recording.

□ □ □

In the mythic universe it was the gods who ruled. In the Platonic system it was the Mind. The Mind, or cosmic demiurge, operated according to a universal plan fully known to the human soul before inhabiting the physical body. The plan was comprehensible to the soul or mind of each individual and discoverable by inwardly directed inquiry. Plato had unshakeable faith in the existence of an ultimate reality of rational design beyond the shadowy world of confused physical forms. Outward events were mere insubstantial shadows, inward ideas the abiding concrete facts. To this day we are the bewildered heirs of Parmenides' topsy-turvy argument as elaborated by Plato. Experiences are deceptive appearances, conceived ideas the true realities. The scientists, said Plato, who waste their time studying objective phenomena, are deaf to the voice of the soul and have lost themselves pursuing the perishable forms of the imperfect phenomenal world.

In the Aristotelian universe it was the Ideas that ruled, permeating the phenomenal world and making all things rational. Aristotle partially restored the spirits of the age of magic. Transformed beyond recognition, in the guise of Ideas, the spirits in the Aristotelian universe became the innate theoretical properties of the observed world. In this form the spirits are with us still, masquerading as gravitational and other fields, electric and magnetic forces, wavefunctions, potentials, inertia, momenta, energies, pressures, and the rest, disciplined by numerical Pythagorean harmony and obedient to laws more exacting than the social codes of the early people.

□ □ □

The Pythagoreans declared the heavens to be eternal, perfect, and divine. From these premises they drew the conclusion that all heavenly motions were essentially uniform and circular. The problem at the time of Plato was to "save the phenomena." Given the necessity of perfect circular motion for the planets, as seemed reasonable and in accord with the divine plan, how might such motion explain the observations? The planets wheeled in strange paths across the sky, moving sometimes forward and sometimes backward, and even the intricate many-sphered system could not reproduce their motions properly.

Aristarchus of Samos in the next century—the third B.C.—inspired by a germinal Pythagorean idea, showed how the apparent seesaw motion of the planets could be explained. If we assume that the planets, including the Earth, revolve around the Sun, he said, then all other planets as seen from Earth will exhibit the observed forward and backward motion. This heliocentric or Sun-centered system was not accepted; it robbed the Earth of its central location and thereby ran counter to the firmly entrenched anthropocentrism of the mythic universe.

Hipparchus of the second century B.C., reputed to be the foremost of Greek astronomers, built his observatory on the island of Rhodes. By careful observations, and with the aid of epicyclic and eccentric orbits, he was able at last to save the phenomena. The etheric spheres of Aristotle were not just translucent surfaces, but zones of considerable thickness, inside which the planets revolved about centers moving with the spheres. Imagine yourself standing in the center of a circle, and that a boy walks around on the circle, whirling above his head a light attached to the end of a piece of string. The light has epicyclic motion, moving forward and backward as it advances with the boy. If you now stand not exactly in the center of the circle, the light moves with epicyclic motion on an eccentric orbit. By such geometric devices, Hipparchus explained diagramatically the motions of the planets. The origin of the idea of epicyclic motion is unknown. Evidence indicates that epicycles and eccentric orbits were known at the time of Plato and were probably introduced earlier by the Pythagoreans.

Alexandria, a city on the Mediterranean coast of Egypt founded by Alexander the Great, became the cultural and scientific center of the Hellenistic world. If the Academy was the forerunner of the learned society, then the Museum of Alexandria with its research projects and great library was the archetype of the modern institute of advanced learning. Claudius Ptolemy in the second century A.D., at the Museum of Alexandria, served astronomy in much the same way that Euclid, also at the Museum, served geometry four centuries earlier. In his astronomical work, known later to the Arabs as *Almagest* (meaning "The Greatest"), Ptolemy showed by arduous computations how to reconcile the astronomical observations with the machinery of geocentric geometry. (In addition to epicycles he used equants, and this means that the boy, whirling a light around above his head, walks on an eccentric circle with uniform angular motion as seen from the point where you stand.)

The Aristotelian universe had become a geometric marvel, a world of physical and etheric harmony, regulated by principles intelligible to the human mind. It lacked only a secure place for the supreme being of the emergent monotheistic religions, and this defect was remedied by the Jews, Arabs, and Christians in the medieval universe.

But it failed to incorporate many important scientific ideas. It was an Earth-centered universe of finite size, and its advocates had no recourse other than to reject outright the thought of a boundless universe of countless worlds. The atomist universe, in any event, was atheistic and, like most unpleasant things, quickly forgotten. The god-defying idea that the Earth and heavens were of similar substances and obeyed similar laws, as taught by Anaxagoras, was repugnant and unacceptable for reasons obvious to all worshipers of heavenly gods. The Milky Way stretching across the night sky was thought to be an atmospheric phenomenon, perhaps the burning of marsh gas, and this idea naturally took precedence over the profane suggestion by Democritus that the Milky Way might be an aggregation of distant stars. The heretical idea that the Earth rotates, and not the sacred heavens, advanced by Heracleides of the fourth century B.C., seemed too absurd for words. The heliocentric idea proposed by Aristarchus, though accepted by Archimedes, seemed too blasphemous to be discussed and taken seriously.

Ptolemy repeated for the benefit of his readers the old commonsense arguments demonstrating that the Earth necessarily is immovable and irrotational. When we jump upward off the ground, we fall back on the same spot, and this well-known fact, which even the gods are powerless to alter, wrote Prolemy, proves that the Earth is fixed and does not rotate. For if the Earth's surface moved, we would fall back on a different spot, owing to the movement of the surface while out of contact with it. Next time you fly in a plane, toss a coin in the air, and note how it falls back into the palm of the hand. According to Ptolemy's argument the released coin should fly backward and strike the rear of the cabin, owing to the motion of the plane. Ptolemy might at least have dropped stones from the mast of a moving ship and noticed how they fall to the foot of the mast, not elsewhere, but this experiment had to await Pierre Gassendi in the seventeenth century.

With hindsight we always realize that every universe is falsifiable (meaning it can be tested and shown to be wrong), and fortunately for our peace of mind we see clearly only how to verify, not falsify, our own particular universe (as discussed in *The Witch Universe*).

The brilliance of Greek science waned and finally died at the end of the second century A.D. Galen's anatomy and physiology, the mathematics of Diophantus, and the Ptolemaic system were among its last conspicuous achievements.

The intellectual giants of ancient Greece created cosmic systems that have since transformed the outlook of almost all human beings. They turned the tide against the mythic universe and reactivated the world in ways that puzzle us to this day.

Some historians tell us that Thales was the first to think that all nature was comprehensible to the human mind. But I think they are wrong, that Thales was not the first by any means. He gave back to nature its comprehensibility that long ago had been taken away and given to the gods. We may think it odd, even inane, that Thales said all things come from water; the important point is that he said water and not the gods.

Consider what might have happened if Thales had not lived. Only a small fraction of all people now inhabiting the globe would be alive, and most of those living would probably be serfs and slaves in a mythic universe, governed by divinely appointed despots, with premature death by disease and warfare the common lot. If this seems too much of an exaggeration, then throw in the rest of the Ionian scientists, and include the Pythagoreans for good measure. Little doubt then remains that the present-day world would be vastly different and much less pleasant than the modern physical universe.

Should science decline in the future and its mental temper vanish, there will not, I fear, be a renascent age of magic or a revived age of romantic chivalry. More probably, it will be a pre-Hellenic barbaric world of dead matter, ruled by gods forever commissioning wars, kings forever beating the war drums, and priests forever blessing the spears; a mythic universe in which only inordinate courage can search and only genius can find the hidden doorway to a new and harmonious universe of awakened matter.

Of all the miracles of the mythic universe the most remarkable and unlikely was the emergence of science.

5

The Medieval Universe

"I drew these tides of men into my hands and wrote my will across the sky in stars," wrote Lawrence of Arabia in *The Seven Pillars of Wisdom*.

Human tides have washed across the globe, crushing nations and carving out empires, led by god-inspired men who sought to write their will across the sky in stars. One such leader was Alexander the Great, who crossed the Hellespont in the fourth century B.C., subjugated Asia Minor and Egypt, vanquished the armies of the Persian Empire with his invincible cohorts, quelled the turbulent forces of Afghanistan, crossed the Hindu Kush, invaded and defeated the nations of the Punjab.

Eastward flowed Hellenic philosophy and science in the wake of Alexander's conquests; westward flowed oriental philosophy and religion. Westward into the Mediterranean world came the Zoroastrian lord of light—the glorious Ahura Mazda—embattled with the lord of darkness, bringing the belief that the soul is divine and the worship of gods other than the appointed one a sin. Westward into the Roman legions came the religion of the dying and resurrected martyred god—the triumphant Mithra—bringing the sacramental eating of the flesh of the god and the notion that evil is the privation of good. Westward came the Babylonian story of the creation and the flood and the Persian stories of heaven and hell, the last day of judgment, and the resurrection of the dead, all of which reshaped the theology of the Greco-Roman world in preparation for the rise of Christianity.

Then, centuries later in the hinterland beyond the Volga a nation of Huns erupted in pandemonium, attacked by fearsome nomadic Avars. Hordes of dislodged Huns swept through the empires of the Ostrogoths and Visigoths. The Goths fled before the storms of arrows and crossed the Danube into the Roman Empire. The fleeing Goths pressed on the Vandals, also a Germanic nation, who joined in the pell-mell rush to escape the tide of terror. Following the death of Attila the Hun—the "Scourge of God"—the Huns were defeated and dispersed by combined Gothic, Celtic, and Roman armies. The crazed Vandals, who had lost their homes, wives, and children, sacked Rome, then again fled before the Goths and set up a kingdom in North Africa from which they harried the Mediterranean with pirate fleets until suppressed by Byzantine forces.

Theodoric the Goth became king of Italy toward the end of the fifth century and sought to restore order amid the ruins of the Roman Empire. According to Edward Gibbon, in his history the *Decline and Fall of the Roman Empire,* the defeat of the empire was the "triumph of Barbarism and Religion." Historians now offer other views: political corruption, military anarchy, economic chaos, bureaucratic oppression, excessive taxation, and breakdown in the judicial system had destroyed the empire from within before the barbarians gained their victories.

In the seventh century the Arabs poured out of their deserts and founded the Islamic Empire that stretched from Spain to India. The message of Islam (meaning "pious submission") was the power and glory of the "One God." Thriving trade by land and sea linked an empire of unusual religious tolerance. Nestorian beliefs that Christ was an inspired human being and his mother an ordinary mortal—not a holy virgin— influenced the formulation of Islamic doctrine; to this day Mohammed is looked upon as the inspired vehicle of the voice of God.

For thousands of years nomadic tribes of Mongolians and Turkic people had periodically sallied forth from the steppes of Central Asia. To withstand their benighted assaults and ravages, the Chinese in the third century B.C. built the Great Wall on their northern frontier. Once more, in the thirteenth century, under the leadership of Genghis Khan, the descendents of the "blue wolf and gray dove" rallied, and again the warrior horsemen swept southward and westward. The earth trembled to the thunder of hoofbeats, and the sky darkened with the sack of cities. The Mongolian Empire of Kublai Khan covered more than a quarter of the land surface of the globe. Along the Silk Road traveled intrepid European adventurers, including the young Marco Polo, who were dazzled by the unsuspected magnificence of oriental civilization.

With the death of Kublai Khan, at a time when Europe stood at the edge of being engulfed, the empire dissolved into a maelstrom of huge contending armies. "A monstrous and inhuman race of men" appeared from the East, according to an Arab ambassador desperately seeking aid from the Europeans. But aid was not forthcoming and in any event would have been to no avail. The old Byzantine Empire of Egypt, Asia Minor, and Balkan Peninsula was overwhelmed and gathered into the fold of the Ottoman Empire, and Constantinople finally succumbed in the fifteenth century. The Byzantine bastion that defended Europe for more than a thousand years had at last fallen.

Out of the holocaust, along caravan routes and in the wake of armies, came disease-infested rats. More than half the populations of Asia, North Africa, and Europe died in the plagues that immediately followed.

In scant words this is the historical background of the medieval universe of the Middle Ages.

◻ ◻ ◻

The medieval universe—the Eternal City and dream of Saint Augustine—reached its zenith in the high Middle Ages of the twelfth to fourteenth centuries and was the last and grandest of the mythic universes.

The religious rudiments of the medieval universe consisted of Hebraic scriptures and gospels. The history of the world had unfolded according to a divine plan whose major events comprised the creation, fall, flood, election of the Israelites, exodus from Egypt, works of the prophets, exile in Babylon, incarnation, crucifixion, and resurrection. All other incidents, such as the Egyptian and Mesopotamian civilizations, Roman Empire, and conversion of Constantine, served in the implementation of the plan.

Man and woman in the beginning were created perfect, but because of their original sin of willful disobedience, they fell from grace into a state of spiritual deprivation. God sent his only son, the Redeemer, to show the way of atonement and salvation. The wrath of God could be averted by appeal to his mercy, but the original sin remained forever unforgivable until the last day of judgment, when all persons would receive their just deserts: the evil condemned to everlasting torment, the good restored at last to the spiritual grace of the first man and woman.

◻ ◻ ◻

It was inconsistent with doctrine that a Christian should live in slavery, possessed body and soul by a human master rather than by God, and with the spread of Christianity into Europe in the early Middle Ages, slavery retreated and almost vanished.

The Benedictine monks in the sixth century launched a missionary enterprise and established Western European monasteries and schools. The Benedictines taught the elements of orthodox doctrine and also the trivium, consisting of grammar, logic, and rhetoric, which in earlier centuries had formed the basis of the curriculum in the Roman schools. After Charlemagne the monastic schools taught in addition the quadrivium, consisting of arithmetic, astronomy, geometry, and music; the trivium and quadrivium together formed the liberal arts. Compilations by Roman encyclopedists, such as Pliny's *Natural History*, served as supplementary texts in the schools. The works of Boethius, a renowned scholar in the early Middle Ages whom Theodoric executed on a charge of conspiracy, were indispensable in the education of Europe. During his imprisonment while awaiting execution, Boethius wrote the *Consolation of Philosophy*, and this remarkable work and his fragmentary translations of Euclid, Aristotle, and Ptolemy with commentories were of lasting value. By fostering devout learning the Church succeeded in creating a European cultural unity.

Meanwhile, under the rule of the caliphs, the arts, crafts, and sciences waxed luxuriantly. Greek, Jewish, Persian, and Indian scholars flocked to centers of learning in Baghdad, Damascus, Cairo, and Cordoba, where libraries were stocked with ancient manuscripts. Novel lifestyles, ideas, and technologies filtered into Western Europe, and slumbering Europe, bestirred by this vigorous civilization, slowly awoke.

The stirrup was the key that unlocked European feudalism and created an aristocracy of mounted warriors in an age of chivalry. Legends are replete with accounts of damsels in distress but remiss in not mentioning the highly skilled smiths who manufactured and maintained the knightly armor. The heavy wheeled-plough, padded horse-collar, and nailed horseshoe revolutionized agriculture and increased the production of food. The standard of living rose, the population grew, and barbarian vernaculars of Latin evolved into romance languages of French, Italian, Spanish, Portuguese, and Rumanian.

With slavery banished, Europe threw off the traits of ancient living, and mind-numbing harsh toil disappeared in a society sustained by the skills of artisans and the investments of financiers. The Cistercian monks with their mechanized communities and labor-saving methods set the example and pioneered the technological revolution. Rivers, winds, and tides were harnessed to supply power to water wheels, windmills, and tidal mills. Mechanical contrivances began to abound, either invented or copied (many from the inventive Chinese), and transmission shafts, driving belts, gear trains, flywheels, cranks, cams, treadles, springs, and spinning wheels became commonplace throughout a mechanical-crazy Western Europe. Mills everywhere in the high Middle Ages busily ground, mixed, crushed, tanned, fulled, sawed, pulped (for paper making), and also operated bellows and trip hammers for the forging of iron. It became an age that built the great cathedrals.

Mechanical clocks—the products of advanced technology—appeared in the late thirteenth century. In *Medieval Technology and Social Change,* Lynn White writes,

> Somethings of the civic pride which earlier had expended itself in cathedral-building was now diverted to the construction of astronomical clocks of astounding intricacy and elaboration. No European community felt able to hold up its head unless in its midst the planets wheeled in cycles and epicycles, while angels trumpeted, cocks crew, and apostles, kings, and prophets marched and countermarched at the booming of the hours.

And of course we must not forget the compass, nor the invention of spectacles that in this age of genius extended the working life of scholars and craftsmen.

A revolution had occurred, unique in history, in which specialized skills no longer remained confined to palaces for the entertainment of royalty, but became widespread for the benefit of all in a mechanized society. The Middle Ages (long referred to as the Dark Ages by historians in the liberal arts who despised the "servile" arts) were a time of social and intellectual change of immense significance, in which technology and then science gathered momentum and wrenched society from the lifestyles of the ancient world and created new outlooks and ways of living. Ordinary people, skillful and industrious, released from soul-destroying drudgery, discovered they had social value.

□ □ □

Already by the end of the ninth century Western Europeans knew the Earth was a sphere and realized that their universe, contrary to Hebraic myths, had geocentric symmetry. Inquisitive monks thirsting after new knowledge began to take an avid interest in the legacy of classical antiquity. The cause of this interest was undoubtedly Islam with its foothold in Spain and Sicily. Arab manuscripts, when translated into Latin, fomented intellectual unrest, and tantalizing fragments of Euclid, Aristotle, and Ptolemy triggered trains of novel thought.

An age of translations commenced. New words of Arabic origin gained currency, such as algebra, alkali, azure, camphor, cipher, borax, elixir, jasmine, jute, saffron, sherbert, zenith, and zero. Numerous scribes in the twelfth and thirteenth centuries industriously translated into Latin whatever Greek documents they could lay their hands on.

The flood of new knowledge burst the bounds of the monastery and cathedral schools, and much of it remained in the hands of translators and their successors who turned into professional teachers. Communities of learned educators at Salerno, Bologna, and Padua taught the liberal arts, medicine, and law. These communities were the early universities to which students traveled from far and wide. Knowledge and learning had become the surest route to high office. Students paid their fees to the professors and formed unions to ensure they got their money's worth; the professors in response formed their own academic unions, or faculties, which regulated the dispensation of degrees and doctorates (licenses to teach). Much the same as today, admission to the faculty depended on qualifications and letters of recommendation.

The universities of France and England developed a formal structure and functioned under royal charter with papal sanction. Students were subject to canon law and exempt from common law. Colleges, or endowed halls of residence for the less wealthy, promulgated rules of decorous behavior and were a conspicuous feature of the universities at Oxford and Cambridge.

Control at the University of Paris was vested in the chancellor, a dignitary of the Church, and the faculties of theology, medicine, law, and arts had each a presiding dean. The faculty of arts, the largest, taught the trivium and quadrivium, both greatly enlarged by the influx of new knowledge. The curriculum at Paris in the mid-thirteenth century included courses on astronomy, meteorology, physics, animals, plants, causes, ethics, economics, sense and sensibles, sleep and waking, memory and remembering, life and death, and the soul. Students worked hard and were possibly better educated than today. A master's degree in arts took usually 6 years of study, followed by 8 more years of study for those aspiring to a doctorate in theology. Of the approximately seventy universities scattered around Europe in the late Middle Ages, most were patterned on the Parisian model, with theology as the supreme subject of study. Charles Haskins in *The Rise of the Universities* remarks, "We are the heirs and successors not of Athens and Alexandria, but of Paris and Bologna."

Hitherto Roman law had remained entangled in Gothic codes. With the revival of classic learning came the study and practice of Roman law, leading to the restoration of judicial torture as an accepted means of determining guilt and innocence. In the witch-craze of the Renaissance, hundreds of thousands of victims were tortured in accordance with the principles of Roman judicial inquiry. Canon law was preferred to Roman law in English universities, and in England, after Magna Carta, torture was generally extra-judicial and officially reserved for acts of treason, consistent with Gothic tradition.

At first the universities were dominated by the mendicant orders—Franciscans and Dominicans—whose members ranked among the most learned thinkers of the Middle Ages. The Franciscan monk Roger Bacon typified the fluidity of thought of this period. Wholeheartedly he embraced Aristotelian empirical science, and with his inventive mind he sought to unravel the secrets of nature. More than anyone else he foresaw the outcome of the technological revolution. He predicted ships that would move without sails or rowers, vessels capable of exploring the bottom of the seas, flying machines, and he prophesied, "wagons may be built which will move with incredible speed and without the aid of beasts."

☐   ☐   ☐

The most influential of the new ideas in the universities came from the works of Aristotle; they created intellectual excitement, opening the door into a world that exalted the individual as a rational being.

Averroes, an Arab scholar of Cordoba in Spain, showed how Aristotelian knowledge could be adjusted and harmonized with Islamic beliefs; Moses Maimonides, a learned rabbi also of Cordoba, did much the same for Judaic beliefs. Later, in the thirteenth century, Thomas Aquinas, a black-robed Dominican, followed their example and demonstrated how Christianity could be accommodated within a modified Aristotelian framework. Aquinas and other learned scholars took the greatest of contemporary themes, the story of sin and salvation, and wove it into the fabric of the geometric marvel. From their work emerged the medieval universe in its final form.

According to Genesis, "In the beginning God created the heavens and the earth," and unlike the Aristotelian system, the medieval universe had a beginning, therefore a finite age, and was created by God to serve specific ends. The medieval universe retained anthropocentrism, but it was regulated by an anthropomorphic angelology rather than Aristotelian Ideas.

Beyond the sphere of fixed stars lay the primum mobile, a primary sphere introduced by the Arabs, which was maintained in constant motion by divine will; beyond the primum mobile extended the empyrean, a realm of purest fire conceived by Saint Anselm, where God dwelt. The celestial heavens, stratified with planetary spheres, accommodated a hierarchy of angelic beings whose degree of divinity increased with altitude. Aerial and daemonic beings trod a less-orderly measure in the sublunar sphere. Earth had the distinction of being the home of mortal life, and its earthly elements constituted the perishable vessel of the immortal human soul.

Hell, located in the bowels of the Earth, was the abode and source of evil, where with ineluctable certitude wicked people would go to be tormented for their sins. Above the Earth's surface and beneath the sphere of the Moon lay purgatory, where spirits must be purged before ascending further. Guarded by angels, the lunar sphere served as the entrance to the higher spheres of heaven. And beyond the celestial spheres, above the primum mobile, existed the glorious empyrean where God dwelt and looked down and watched over his creation. By compromise and adjustment the learned fathers combined reason with faith and gained a universe of monumental elegance.

Aquinas in an Age of Faith used reason to justify faith. Voltaire in an Age of Reason, half a millennium later, used faith to justify reason. The difference between Aquinas and Voltaire is less than is often supposed. Carl Becker in *The Heavenly City* writes, "What they had in common was the profound conviction that their beliefs could be reasonably demonstrated." Both believed they lived in a meaningful universe of rational design. We live nowadays in a universe where the question of its meaningfulness is itself without meaning. Reason in faith has gone, and faith in reason rests on shaky ground.

The medieval universe was saturated with meaning. It placed human beings in the most conspicuous of all places, at the center of events, on a stage with the spotlight beamed on them as the leading actors in a drama of cosmic proportions. Blessed by religion, rationalized by philosophy, and verified by geocentric science, the medieval universe gave meaning and purpose to life on Earth. Most persons then living could grasp the essentials of the medieval universe and felt impelled to worship its Creator. "Other ages have not had a Model so universally accepted as theirs, so imaginable, so satisfying to the imagination," writes C. S. Lewis in *The Discarded Image*.

□　　□　　□

Many of us live in cities or towns where the night sky is lost in a glare cast by electric lights. Even when the night sky is seen clearly, we glance at it casually, for it means little to us. If we think about it, we know we look out to vast distances in a universe that is dark and mostly void. This was not so for the people who lived in the medieval universe. They looked up unhindered by the glare of electric light to a mysterious and divine panorama of immediate significance, resonant with the choirs of heaven.

It was a radiant universe overflowing with the bright blue light of heaven. The "bright blue firmament" in the Middle Ages actually existed, and the blueness of the daytime sky was not a mere atmospheric phenomenon of scattered sunlight, but the etherial radiance of heaven bathing the Earth. The higher the celestial sphere, the more dazzling became the supernal light. At nighttime, according to medieval scholars,

The Lord creates and maintains the universe. From Martin Luther's *Biblia*, published by Hans Lufft, Wittenberg, 1534.

the blue light of heaven could not penetrate the Earth's shadow. Demons from nether regions arose at night and roamed freely in the darkness. The alternation of day and night gave testimony of the unending struggle between the lords of light and darkness, the powers of good and evil.

To people of those times the magisterial medieval universe seemed immense in size. Lewis writes in *The Discarded Image*, "And the fact that the height of the stars in medieval astronomy is very small compared with their distance in the modern, will turn out not to have the kind of importance you anticipated. For thought and imagination, ten million miles and a thousand million are much the same."

In the modern physical universe the Earth is extremely small, but so is everything else, including the galaxies. The medieval universe, with its outer boundary at finite distance, provided a standard of comparison that made the Earth's smallness vividly apparent. "To look up at the towering medieval universe," says Lewis, "is much more like looking at a great building. The 'space' in modern astronomy may arouse terror, or bewilderment, or vague reverie; the space of the old presents us with an object in which the mind can rest, overwhelming in greatness, but satisfying in its harmony."

Universes for more than 100,000 years have progressively grown larger. Always, the preceding known universes must have seemed comparatively small, yet to their inhabitants they were invariably large. Amazement at the extent of the modern universe is not a new sensation, for it repeats the constant amazement of the past and is probably as old as cosmology. Any universe that encompasses all known things seems vast and causes amazement.

The fantasy of journeying away from the Earth as a space traveler, ascending through the successive spheres, and then looking back and seeing the Earth as a distant and tiny orb had been known since the first century B.C. from the writings of Cicero. It was an imaginary device often employed in the Middle Ages as a means of emphasizing the grandeur of the heavens and the relative smallness of the Earth. Dante employed it with great effect.

□ □ □

The Neoplatonic idea of God at "the center of the world," elaborated in the mystical writings of Pseudo-Dionysius (an unknown author of the fourth or fifth century), never entered the mainstream of Christian doctrine. Unlike the geocentric classical picture, the Pseudo-Dionysian universe had inverted structure and was theocentric. God occupied the center of the universe, as seemed proper and fitting to the Gnostics and

Neoplatonists, surrounded by angelic spheres of increasing size; beyond the outermost sphere lay darkness, where human beings appropriately dwelt.

The theocentric universe was too obviously modeled on ignorant pagan beliefs to gain wide acceptance; even the most highbrow clerics in their ivory towers could not brush aside the astronomical fact that the Earth and not God had central location in a universe of rotating celestial spheres. Christianity and Mohammedanism were both nurtured on the Platonic concept of God as omniscient and omnipresent Mind, and neither religion could accept the thought of God as confined to a single and fixed point. In the famous throne verse of the Koran we are told, "He knows what is before and behind men. They can grasp only that part of His knowledge which He wills. His throne is as wide as heaven and earth, and the preservation of them wearies Him not, and He is the Exalted, the Immense."

The ingenious Dante Alighieri in the early fourteenth century, with artistic licence, succeeded in bringing together within a unified scheme the geocentric and theocentric systems. In the *Divine Comedy* ("divine" was added later, and "comedy" means a happy ending) Dante placed the angelic spheres within the empyrean in such a way that the celestial and angelic spheres acted as mirror images of each other.

It is easy to construct a simple model that illustrates Dante's universe. Take a dinner plate, or a large disk of white cardboard, and on one side mark in the center a point to indicate the Earth. Draw now around this point eight concentric circles of increasing size to represent the celestial spheres (Moon, Mercury, Venus, Sun, Mars, Jupiter, Saturn, and Stars), and let the rim of the plate be the primum mobile. On the other side mark in the center a point to indicate God. Around this center again draw eight concentric circles of increasing size to represent the angelic spheres (Seraphim, Cherubim, Thrones, Dominions, Virtues, Powers, Principalities, and Archangels), and let the rim of the plate in this case be the sphere of the Angels. On one side of the plate we have the geo-centric world of celestial spheres, on the other side the theocentric world of angelic spheres, and mediating between the two, at the rim, are the Angels occupying the primum mobile. This model in which Earth and God are the antipodes of a symmetric universe shows how Dante brought into harmony two totally dissimilar systems.

In his imaginary journey, as recounted in Canto 28 of *Paradise*, Dante leaves the Earth and ascends through the celestial spheres to the rim, and he sees on the other side of the universe a panoramic view of heaven:

> *One point I saw, so radiant bright,*
> *So searing to the eyes it strikes upon,*
> *They needs must close before the searing light.*

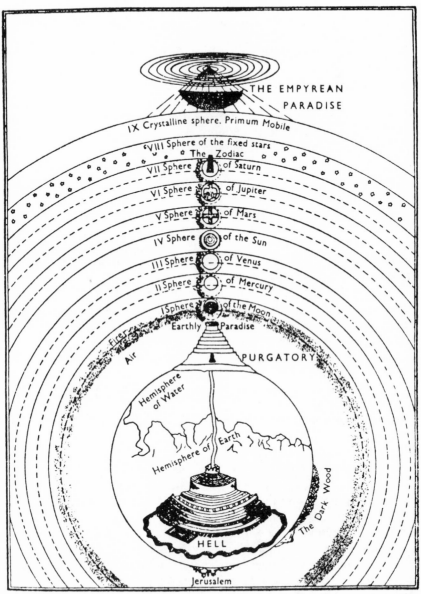

Dante's conception of the universe, in which the angelic spheres occupy the empyrean. This popular engraving depicts a disproportionately large Earth and fails to show the angelic spheres on the same scale as the celestial spheres.

. . . .
About this point a fiery circle whirled,
With such rapidity it had outraced
The swiftest sphere revolving round the world.

. . . .
This by another was embraced,
This by a third, which yet a fourth enclosed;
Round this a fifth, round this a sixth I traced.

. . . and so on. While standing at the rim he sees before him a brighter world similar in structure to the one left behind.

Dante's synthesis made little impact on cosmology. The notion that God could be geometrized, while permissible in metaphorical flights of poetic fancy, was generally unacceptable, for it presupposed that the form and nature of God could be apprehended by the human mind. The standard model, to which Dante subscribed in his other works, consisted of a set of angelic spheres superposed upon a corresponding set of celestial spheres. The angels, like Aristotle's Ideas, were distributed throughout the visible heavens and provided the motive power that maintained the ceaseless rotations of the celestial spheres. The Angels occupied the sphere of the Moon, the Archangels the sphere of Mercury, and so forth, to the Seraphim who occupied the primum mobile and were nearest to God.

□ □ □

Most astronomers from the Babylonians to the Newtonians regarded themselves as astrologers. It must be understood that *astrology* had not its present meaning. As the name implies, it was the science of planets and stars, of celestial emanations and influences, dealing with the interactions, physical and etheric, of celestial bodies and the Earth. Eclipses, tides, and seasons of the year were associated with astral forces and therefore of astrological interest. Astrology as a science has since been superseded by astronomy, and nowadays astrology concerns itself with the effect of planets on the mundanities of human life.

After Alexander the Great had opened the floodgates to oriental cults and mystery religions, astrology became linked with astrolatry (worship of celestial bodies) and the devious temple arts of astromancy. Astromancy, consisting of astral divination and soothsaying, reached a climax during the reign of the imperial emperors of Rome. Thereafter it declined in the Middle Ages with the spread of Christianity. Its pagan overtones of skygod veneration had no place in a universe where all was governed by the will of God. Observations and studies of the night sky fell into disrepute because of the association of astromancy with astrology.

Geoffrey Chaucer's *Treatise of the Astrolabe,* dealing with celestial observations, was for a long time referred to as an astrological work. In the late Middle Ages and until the sixteenth century astrology meant literally the science of the planets and stars; astronomy, so far as it had different meaning, consisted of the naming and identifying of stars and constellations. Astrology has since advanced and developed into the modern science of astronomy. It is a pity that astrology was not retained as the proper name for astronomy; we have biology and do not use bionomy as a preferred substitute.

Universities still receive letters addressed to the "department of astrology," and astronomers nowadays are continually surprised by the number of persons unable to tell the difference between astronomy and astrology. What nowadays is called astrology is nothing more than a mythical residue of astromancy and astrolatry, and casters of horoscopes are astromancers practicing at the level of pagan soothsayers. The medieval universe in the late Middle Ages and the Renaissance gave full rein to fanciful superstitions of every kind imaginable. With the resurgence of demonology, necromancy, witchcraft, came also astromancy; all have gone, save astromancy, which hides behind the once-respectable name of astrology.

In the first half of the sixteenth century the Swiss physician Paracelsus and the Flemish anatomist Vesalius discarded the mythically encrusted medical lores of Galen and Avicenna and laid the foundations of modern medicine; at about the same time Copernicus discarded the Ptolemaic system and began a new age in astronomy. Chemistry, divorced from alchemical and medical arcane lore, began as a natural science with Robert Boyle's *Skeptical Chemist* in the mid-seventeenth century.

The medieval universe from yet another viewpoint was the Great Chain of Being. The Neoplatonists developed the notion that the organic world consisted of countless graduated species of living creatures. This view of the living world later became popular in the late Middle Ages and influenced European thought up until the nineteenth century. Link by link, the great chain of sequential lifeforms descended from human beings through beasts and plants to insensible matter and ascended through angelic beings to the throne of God. Human beings were the central link connecting the angelic and brute realms. All known and imaginary species fitted into the grand arrangement, and no gaps could exist to blemish its perfect creation and continuity.

Arthur Lovejoy recounts in *The Great Chain of Being* how this theme captured the imagination of Europeans and made an indelible impression on their literature and art. In his *Essay on Man*, Alexander Pope wrote:

> *Vast chain of being! which from God began,*
> *Nature aetherial, human, angel, man.*
> *Beast, bird, fish, insect, what no eye can see,*
> *No glass can reach; from Infinite to thee,*
> *From thee to nothing.—On supreme powers*
> *Were we to press, inferior might on ours;*
> *Or in the full creation leave a void,*
> *Where, one step broken, the great scale's destroyed;*
> *From Nature's chain whatever link you strike,*
> *Tenth, or ten thousandth, breaks the chain alike.*

Lyrics voiced scientific beliefs. The connecting links forged by the Creator disallowed any possibility of evolutionary change. Were only one species to change or disappear, the severed chain would crash to the ground. This grand picture of a biological chain of immutable forms was what the evolutionists had to fight against in the nineteenth century. The chain interconnected the universe, moored the living world to God, and secured for human beings a central position of semi-divine importance.

The Middle Ages made articulate yet another concept, the principle of plenitude, implicit in Judaic and Christian doctrine. The notion of plenitude sprang from the belief that a benevolent Creator had given to human beings a richly endowed Earth for their own profit and exploitation. This sentiment is expressed in the Eighth Psalm: "Thou has made him a little lower than the angels, and hast crowned him with glory and honour. Thou madest him to have dominion over the works of thy hands; thou hast put all *things* under his feet," sheep, oxen, beasts of the field, fowl of the air, and fish in the sea, for the benefit of mankind.

The principle of plenitude dovetailed into the Great Chain of Being. The Earth possessed unlimited wealth of every kind and displayed in profusion all possible forms of graduated life, with no gaps or missing links in the great chain. Land, sea, and air necessarily teemed with living creatures in inexhaustible supply, and depletion to the point of extinction was quite impossible, for missing links (a pre-Darwinian expression) would imply a makeshift creation. To doubt plenitude was equivalent to denying God's munificence, and not consume to the utmost whenever possible was equivalent to rejecting God's gifts. Belief in plenitude and the divine right to squander drove European nations to the extreme limits of exertion; civilizations everywhere toppled before their fierce hunger, and their imperial empires of merchant adventurers girdled the globe.

The myth of plenitude, which lies at the roots of modern political ideology, was the motivating philosophy, and whenever anything showed signs of extinction, hunters, trappers, whalers, fishermen, lumbermen, miners, farmers, explorers, fortune-seekers, politicians, clergymen, and the consuming public were overtaken by amazement and unable to believe that exhaustion had actually occurred.

□　　□　　□

Already in the middle of the thirteenth century alarmed ecclesiastics were expressing concern that the conciliation of Christianity with Aristotelianism had gone much too far. The geocentric universe was all very fine, but if it placed constraints of any kind on the power of God, or if it meant that God existed only outside the primum mobile, or that God could not move the Earth if he so willed, or could not create worlds other than the Earth, then it was false and contrary to orthodox belief.

Etienne Tempier, Bishop of Paris, issued in 1277 the famous, some say infamous, 219 condemnations anathematizing "the execrable errors which certain members of the faculty of Arts have the temerity to study and discuss in the schools." All ideas tending to restrain the power of God were roundly condemned, for God was to be apprehended by faith and not intellectual arrogance. God has unlimited power, said Tempier, and hence it is inadmissable to suppose that God is defined and circumscribed by boundaries. The empyrean may indeed be the realm closest to the throne of God, as Anselm suggested, but it must be realized that God dwells everywhere and is not limited in any way by the fictitious necessities of philosophy. The bishop's strictures put a stop to all speculations that might appear to diminish the Christian concept of God, and one consequence of this dampening of enthusiasm was the coolness the *Divine Comedy* received when published a few decades later.

The condemnations of 1277 are a landmark in the history of cosmology. By reasserting that God is neither here nor there, but everywhere, they redirected inquiry and prepared the way for the Copernican Revolution.

In the years to come, the universality of God became more and more reflected in the attributes of the created universe. If God indeed was boundless, why should not the universe—the handiwork of God—also be boundless? If God was without specific location, then why should human beings claim central location, and if God had unlimited power, why should not the Earth move if he so willed? Theological concepts concerning the universality of God inspired secular concepts concerning the nature of the universe. What was potential in the medieval universe became actual in the Newtonian universe.

The medieval universe has gone, leaving us clinging to its customs and conventions. In our social institutions, languages, and familiar topics of everyday conversation; in our domestic life; in our technologies, labor-saving practices, and numbers; in our manners, greetings, letters, nursery rhymes, and superstitions; in our ceremonies, rites of passage, acts of charity, notions of decency, and expressions of love can be found the enduring remnants of the medieval universe.

Though we feel lost in the modern and seemingly meaningless physical universe, deep down in our world pictures we find comfort in hereditary beliefs and think as medieval people. In the West we hold to the values of the medieval universe that lasted for a thousand years.

6

The Infinite Universe

Etienne Tempier roundly condemned all who dared to trifle with the power of the supreme being. Scholars were free to admit reason into faith and could debate and teach on sundry matters, provided that God always received full recognition as an all-powerful being free of self-contradiction. Here was the Trojan Horse, introduced by the well-intentioned bishop, from which sallied thoughts in years to come that finally overthrew the medieval universe.

☐ ☐ ☐

Professors at Oxford and Paris in the fourteenth century made great progress in clarifying the nature of space, time, and motion. William Heytesbury and his colleagues at Merton College defined velocity and acceleration and then succeeded in calculating the distance traveled in an interval of time by a body in constant acceleration. William of Ockham participated in these studies while fighting a battle against needless abstractions. His celebrated principle of theoretical parsimony—known as Ockham's razor—states that in the employment of ideas, "it is foolish to accomplish with a greater number what can be done with fewer."

Jean Buridan, a professor at Paris and formerly Ockham's student, revived the notion of impetus that can be traced back to Hipparchus in the second century B.C. and is now referred to as momentum. According to Buridan, impetus is proportional to the velocity of a body and also its quantity of matter (now referred to as mass), and the impetus of a thrown body maintains the body in a state of motion. Aristotle, lacking the notion of impetus, supposed that rest is the natural state of all bodies, and a moving body must be pushed or pulled constantly by a force and come to rest immediately when the force ceases. Walking and swimming are examples of motion in a resistive medium in which movement is maintained by continual effort. But the planets move freely, for they encounter no resistance, argued Buridan, and therefore need not be continually pushed or pulled along in their orbits by angelic forces. Instead, they move of their own accord because of the impetus imparted to them by God in the beginning.

Buridan promoted the idea that falling bodies accelerate similarly, and when bodies of different weights are dropped side by side, they reach the ground simultaneously. This idea had been mooted by Philoponus of Alexandria in the sixth century, and after its revival by Buridan was adopted by other scholars, such as Leonardo da Vinci, and successfully tested by Simon Stevin of Belgium in the sixteenth century. Allowance must be made for the fact that air resistance introduces complications; feathers, for instance, fall as fast as stones in a vacuum but not in the resistive atmosphere.

From the fifteenth to the seventeenth centuries the universities added little more to the advance of science. They lagged behind, mired in the sticky problems of reconciling Aristotelian doctrine with religious dogma. It was the theologians, Bishop Oresme, Cardinal Cusanus, and Canon Copernicus, working outside the universities who toppled the Ptolemaic system in what is referred to as the Copernican Revolution.

☐   ☐   ☐

Bishop Nicole Oresme of France in the fourteenth century had much to say about the unlimited power of God. "Motion is perceived by ordinary mortals," he said, "only when one body alters its position relative to another." Mind you, mortals' perceive only the relative motion, but God knows always the absolute motion. We observe the heavens turning around the Earth, but only God can tell whether the Earth is still and the heavens revolve, or the Earth rotates and the heavens are still, and neither is beyond the power of God. Our impression that the Earth sits still might easily be wrong, for do not sailors in a moving ship have the impression that their ship is stationary and the sea and shore move? If the Earth moves, then everything on its surface, including the seas, atmosphere, and ourselves, moves with it and shares in its impetus, and the old arguments claiming that the Earth is necessarily at rest are false, and furthermore limit the power of God.

Who knows whether one or many worlds were created? The inhabitants of other worlds, if such worlds exist, said Oresme, may also have the impression that they occupy the center of God's creation.

Oresme likened the universe to a delicately adjusted clock, and in his thoughts we see the first stirrings of a new world in which the geometric marvel of the ancients would be transformed into the clockwork marvel of the Newtonians.

The budding ideas of the bishop flowered in the fifteenth century in the fertile mind of Cardinal Nicholas of Cusa. In his work *Of Learned Ignorance,* the cardinal said, "every direction is relative," and "wherever a man stands he believes himself to be at the center." The ancients formed the opinion that the Earth is stationary and at the center of the universe. But has not God infinite power and wisdom, and apart from what mere mortals think, the Earth may move and not be at the center.

To suppose that the universe is finite with only one center, said the cardinal, when God is infinite and everywhere, is an inference unworthy of God's wisdom and creative power. More likely, "the absolute infinity of God has its counterpart in the infinity of the world as an image," he argued, and consequently "the universe has its center everywhere and its circumference nowhere."

Belief that God was unlimited, and therefore infinite and everywhere, led Nicholas of Cusa to the conclusion that the created universe was correspondingly infinite and its center everywhere. Wherever one stood in the universe, the stars would spread out much the same in all directions. God is unbounded and equally the same everywhere, he said, and the created universe similarly is unbounded and equally the same everywhere.

Nicholas of Cusa, theologian, scientist, and statesman, had an active and inventive mind. He advocated the counting of pulse beats as a diagnostic aid in medicine, developed spectacles for the correction of nearsightedness, arranged jets of water to measure the passage of time, and claimed that plants draw on the atmosphere for nourishment. He hit also on the momentous cosmological argument that if the starry heavens are the same in every direction as seen from the Earth and the universe has its center everywhere, then the starry heavens must appear much the same in every direction as seen from all other places. The logical conclusion was that all places in the universe were much the same; for if God was unbounded and equally everywhere, then the universe itself was unbounded and equally the same everywhere. The hypothesis that "all places are alike" (Einstein's words) is today known as the cosmological principle.

The epic poem *On the Nature of the Universe* by Lucretius, in praise of the infinite atomist universe, was discovered in 1417 hidden away in a monastery. Possibly Nicholas of Cusa was influenced by hearing or reading about this work that contained the essence of his conclusions. Perhaps the "discovery" of the poem was facilitated by Oresme's earlier work, and its adamant atheism precluded open acknowledgment.

Ideas, once born, seemingly have a life of their own. Like germs they breed, spread, mutate, and catch their invaded victims by surprise. When and where an idea originated is often unknown, and many a person believes the idea, as recounted by him, springs unprompted from the recesses of his own mind. A thought drifts as light as thistledown, and sensitive minds responding to its novel vibrations "independently" discover the new idea. A cry from an obscure source echoes with mounting volume and is later attributed to a convenient historical figure. And so with the idea of the infinite universe, which long ago had originated with Anaxagoras; it hung in the air and finally became part of the climate of opinion in the fifteenth and sixteenth centuries.

A brooding discontent in the early sixteenth century was undermining the Ptolemaic system. The cumbersome geometric machinery of epicyclic, eccentric, and equantic movements, so difficult to manage in calendric computations, were clearly inconsonant with the Pythagorean ideal of perfect harmonious motion. The bearings of the geometric marvel were wearing out, and the music of the spheres had acquired a discordant note.

Nicholas Copernicus, an alumnus of the universities of Bologna and Padua and a canon in the cathedral town of Frauenburg, was aware of the dormant heliocentric system of ancient science and had the bright idea that a Sun-centered system might possess simpler movements and greater harmony. For years the canon labored on the computation of heliocentric orbits. We can sympathize with his difficulties; he had to fit the movements of the planets to observations from the Earth, which itself moved as a planet. At last he succeeded in showing that his new system of epicycles worked as well as if not better than the Ptolemaic system. The amazing man had the temerity to believe in the reality of the heliocentric system and did not think it just a convenient fiction devised for the purpose of simplified calculations, as piety demanded.

In his great work, *Revolutions of the Celestial Orbs*, published shortly before his death in 1543 and rivaling in scope the *Almagest*, Copernicus wrote, "Why then do we hesitate to allow the Earth the mobility natural to its spherical shape, instead of supposing that the whole universe . . . is in rotation?" His reasons for supposing that the Earth rotates were much the same as those offered by Oresme. From a rotating Earth it was a short step to the idea of an Earth revolving around the Sun: "We therefore assert that the center of the Earth, carrying the Moon's orbit with

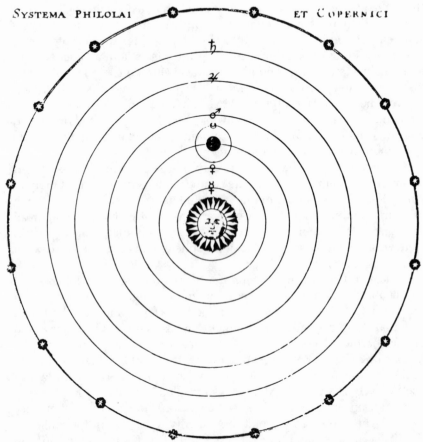

Copernican system with the Sun at the center of the universe, encircled by the planetary orbits and surrounded by the outermost spheres of fixed stars.

it, passes in a great orbit among the other planets in an annual revolution around the Sun; that the Sun is the center of the universe, and that whereas the Sun is at rest, any apparent motion of the Sun can be better explained by motion of the Earth."

In the same year Andreas Vesalius published his great work, *On the Structure of the Human Body*, of equal merit and possibly even greater originality, but of less cosmic significance.

Thirty-three years later the English astronomer Thomas Digges published a popular account of the Copernican system and discarded the outer sphere of fixed stars, which Copernicus had retained. Digges, infected by the bug of infinity, peeled away the outer sphere and distributed the stars throughout endless space: "Of which lightes celestiall, it is to bee thoughte that we onely behoulde sutch as are in the inferioure partes of the same orbe, and as they are hygher, so seeme they of lesse and lesser quantity, even tyll our sighte beinge not able farder to reache or conceyve, the greatest part rest by reason of their wonderfull distance invisible unto us." The stars "devoyd of greefe" and the "habitacle of the elect" occupied the "gloriouse court of ye great god."

The primum mobile had gone, but the empyrean lingered on, and Edmund Spenser in his *Hymn of Heavenly Beauty* rejoiced in the new vision:

> Far above these heavens which here we see
> Be others far exceeding these in light,
> Not bounded nor corrupt, as these same be,
> But infinite in largeness and in height.

The bishop, cardinal, and canon had done their work well.

The fiery Dominican monk, Giordano Bruno, now took to the warpath as the self-elected champion of the infinite universe. In his writings and travels he propagated the message of an endless universe teeming with countless worlds, each an abode of life with its own incarnation, revelation, and redemption. In 1584 he wrote, "Thus is the excellence of God magnified and the greatness of his kingdom made manifest; he is glorified not in one, but in countless suns; not in a single earth, but in a thousand, I say, in an infinity of worlds."

The enraged gods unleashed their hounds, and this "malevolent witch" who dared to oppose the mythic universe of the Renaissance spent his last 7 years shackled in an ecclesiastical prison, was tortured, and in 1600 burned at the stake in Rome.

☐   ☐   ☐

Tycho Brahe, a Danish nobleman of the second half of the sixteenth century, turned to astronomy and made observations of the utmost precision possible before the invention of the telescope. Here at least was order and refuge from the witch universe of the Renaissance. In 1572 a bright light flared in the constellation of Cassiopeia and slowly waned in the following months. The astonished Tycho, after careful observations, found it was indeed a star at great distance. The firmament and its myriad stars was therefore not as perfect and unchanging as everybody supposed. Five years later a great comet appeared, and again by careful observations Tycho showed that the comet could not be a fiery atmospheric phenomenon of the sublunar sphere, for it passed among the planets far beyond the sphere of the Moon. He came reluctantly to the conclusion that because of the motions of comets, the crystalline planetary spheres of Aristotle could not exist.

Tycho agreed with Copernicus that the planets revolve around the Sun, but he compromised by constructing his own system, in which the Earth remained stationary at the center of the universe, and the Sun with its retinue of encircling planets revolved around the Earth.

When news reached Italy of the invention of the telescope in Holland, Galileo Galilei quickly constructed his own telescope and was soon using it for astronomical observations. In 1610 he published a small book, entitled *The Starry Message,* summarizing his discoveries, and on the title page announced:

> The Starry Message—unfolding great and marvelous sights, and proposing them to the attention of everyone, but especially philosophers and astronomers, being such as have been observed by Galileo Galilei, a gentleman of Florence, professor of mathematics in the University of Padua, with the aid of a telescope lately invented by him, respecting the Moon's surface, an innumerable number of fixed stars, the Milky Way, and nebulous stars, but especially respecting the four planets that revolve around the planet Jupiter at different distances and in different periodic times, with amazing velocity, and which, after remaining unknown to everyone up to this day, the author recently discovered.

Galileo believed in the Copernican heliocentric system, and what he saw through his telescope—mountains on the Moon, many hitherto unobserved stars, and the satellites of Jupiter—strengthened his conviction. By observing sunspots he found that the Sun rotates. He noticed that the changing phases of Venus resemble those of the Moon, thus proving that Venus revolves around the Sun. He declared in his forthright manner that to the observant eye and the unprejudiced mind it was obvious that the Earth revolves around the Sun in company with all other planets. But few of his contempories were able and willing to agree with him.

Galileo brought together the strands of medieval thought concerning space, time, and motion, and he repeated the argument that a state of rest is relative and no more natural than a state of motion. Using the idea of impetus (still not clearly defined), he investigated the paths of projectiles; by rolling balls down an inclined surface he confirmed that falling bodies accelerate at a constant rate independent of their weight; he showed that a pendulum swings with a period depending on its length but not the weight of its bob. At the age of 68, in 1632, Galileo published his masterpiece, *Dialogue Concerning the Two Chief World Systems,* in which he contrasted the Ptolemaic and Copernican systems and poured scorn on the physics of Aristotle and the astronomy of Ptolemy. For years his most hostile critics had been academics steeped in Aristotelian doctrine; the clearly heretical character of the *Dialogue* now delivered him into the hands of the Church. Before the steely sauvity of the inquisitors, and confronted in old age with the prospect of torture, he did what any sensible person would have done: on his knees he recanted and abjured the heliocentric system.

Current mythology misrepresents Galileo as a legendary person who miraculously bridged the gulf separating Aristotle and Newton. In fact, Galileo inherited a rich legacy of medieval science, which he analyzed, synthesized, and popularized. He did not perform many of the experiments attributed to him; some were performed by persons whom Galileo failed to acknowledge, and others were imaginary (thought experiments) made possible by his intuitive grasp of scientific principles.

Galileo regarded the circular paths of the planets as God-ordained and in no further need of explanation. He failed to apply mechanical principles to celestial motions and grudged recognition of Kepler's work on this subject.

Living in Germany during the time of Galileo, Johannes Kepler also believed in the Copernican system, and with enthusiasm he adopted the new philosophy calling all in doubt. Kepler agreed with Tycho, "there are no solid spheres," for how could the spheres exist and not be shattered by the passage of comets? Wolfgang Pauli, a modern physicist, argues that Kepler was much influenced by Neoplatonic ideas and regarded the glorious Sun as analogous to the supreme Mind that rightfully occupied the center of the universe. For this reason Kepler vehemently rejected the idea of an infinite universe without center and edge. The very notion of an infinite universe, said Kepler, is terrifying, full of "secret, hidden horror," and "one finds oneself wandering in this immensity in which are denied limits and center, and therefore also all determinate places."

Kepler inherited Tycho's astronomical records and for years strived to explain the movements of the planets within the framework of a Sun-centered system. From this work emerged his three famous laws of planetary motion that served as stepping stones to the mechanistic universe. The first law, the best known and the only one we need mention, states that the planets move in elliptical orbits. After 2000 years astronomy was at last free of its epicyclic fixation.

☐   ☐   ☐

René Descartes in the first half of the seventeenth century helped to clarify the still-murky notions of matter in motion, and he enunciated laws of motion that foreshadowed the work of Newton. By uniting algebra and geometry, thus forming a new branch of mathematics, he forged an essential tool for the construction of the mechanistic universe.

In many fields of inquiry Descartes kept to the beaten track. He echoed Aristotle, "a vacuum is repugnant to reason," for space by itself is sheer nothing, and where there is no distribution of matter, there can be no space. Descartes condemned the atomic theory by arguing that matter must be infinitely divisible. If atoms exist, they exist in a void, with nothing between them, yet a void is ridiculous and repugnant to reason, and therefore atoms cannot exist. Matter exists everywhere; even in what is thought to be a vacuum it spreads out thinly and continuously.

Descartes also rejected astrology, the science of long-range astral forces, and believed that bodies influenced one another only by direct contact. He had no patience whatever with astrological nonsense, and his guiding principle—action by direct contact—swept all its hocus-pocus away. The far-fetched and arcane notion of forces and other influences acting at a distance across empty space was too preposterous to be taken seriously. He pictured the Solar System as a great whirlpool of tenuous fluid in which the planets and satellites were entrained in its vortical motions much like leaves twirling on the surface of eddying water.

Natural philosophers of the seventeenth century, particularly in France, Germany, and the Low Countries, took their cue from Descartes (were thus Cartesians) and would have nothing to do with occult forces of a long-range character. Up until the early decades of the eighteenth century the Cartesians believed that everything was pushed by pressure or pulled by other forces having direct contact, and weird and wonderful were the mechanisms they devised to explain the rise and fall of the tides. Kepler was a notable exception; he did not hesitate to think it possible that the planets were swept along in their courses by a magnetic force emanating from the glorious Sun.

Meanwhile, across the Channel dividing England from France, the magi of a new age were pondering on these matters. Later they would be accused by indignant Cartesians of creating a universe controlled by astrological forces: a universe nonetheless that would eventually enable men to land on the Moon.

☐ ☐ ☐

To educated persons of the Greco-Roman world and the high Middle Ages the weight of a thing was nothing more than its desire to descend to the center of the Earth. The burden of heaviness, called gravity, was the natural penalty of earthly existence. Levity, the opposite of gravity, was the innate desire of less-ponderable things, such as fire, to ascend to more airy and etherial altitudes. The opposing desires of gravity and levity caused all things to seek their proper station in the sublunar realm.

Untutored barbarians experienced great difficulty with the idea of a spherical Earth and could not understand why people on the other side of the globe, standing upside down, did not fall off the surface. But educated persons trained in the rudiments of Greek science had no such difficulty. Wherever a person stood on the surface of the globe the downward direction was toward the center, and gravity was the world-wide desire of all ponderable things to reach that center. The bishop, cardinal, and canon had much to say on the subject of gravity.

Oresme dismissed the notion of levity as superfluous. All things have weight, he said, and whatever rises, such as warm air from a candle flame, is pushed or forced to rise by the descent of what has stronger gravity, such as cold air. Nicholas of Cusa, pursuing Oresme's ideas, surmised that other worlds—created by the infinite power and wisdom of the supreme being—had also centers to which their various parts tended to gravitate. Copernicus in his *Revolutions* expressed the medieval view:

> I think that gravity is nothing more but a certain natural ap-
> petition which the Architect of all things has implanted in the
> individual parts in order that they may unite to attain unity
> and wholeness by adopting a spherical form. It must be as-
> sumed that this property is found even in the Sun, the Moon,
> and other planets in such a way that their observed, un-
> changeable, spherical form is assured.

Gravity until the time of William Gilbert and Kepler remained an inner desire, an "appetition," urging each thing to move downward.

Etheric attractions and repulsions pervaded the medieval universe, reaching across the celestial spheres, eliciting responses from angelic, daemonic, and human souls. The magnetism of lodestones was a convincing illustration of the existence of intangible forces, both astral and esoteric. Magnets were carried as charms to ward off the mischief of demons and witches, and animal magnetism and magnetic power are terms still in use.

William Gilbert, an Elizabethan physician and scientist, became widely known for his learned book on magnetism, published in 1600, in which he showed that the Earth is a huge magnet having north and south magnetic poles. He coined the word electricity, scotched many superstitions about magnetism, showed that garlic could not demagnetize, and assured mariners that their garlic-perfumed breath would not enfeeble their compasses. He conjectured that the planets were attracted to the Sun by an intangible force of a magnetic nature. Kepler was enchanted with the idea.

Gilbert promoted the atomist concept of an infinite universe strewn with countless inhabited worlds. The English at this time were moderately tolerant of their intellectuals, who had the sense to contribute to this happy state of affairs by conceding to the sensitivities of theologians and by acknowledging the handiwork of God in every subject touched upon. Gilbert was the last person to scoff openly at contemporary religious beliefs, and yet his cosmology was as broad in scope as that of Bruno. Instead of being condemned as a heretic he was knighted by the queen and made court physician. His circumspection and willingness to compromise would have been invaluable to the intemperate Bruno and Galileo.

At a time when many natural philosophers in Europe had abandoned the medieval universe, their English colleagues held to the picture of a universe suffused with emanations and spiritual forces. We have only to read the work of Henry More, mentor to Newton, to understand that at Cambridge space was permeated with spirit. The Cartesian belief that space was merely the extension of matter, and by itself could not exist without a supporting distribution of matter, was unacceptable to the English natural philosophers—the Newtonians—for a very good reason.

The Newtonian universe, now in the making, consisted of the old atomist universe with an inlay of medieval spirit. This spirit or quintessence, like matter, had the property of extension. Where there was no matter, pervading spirit alone was capable of supporting the extension of space. The Aristotelian and Cartesian belief that space could not exist where there was no matter, and a vacuum was therefore impossible, denied the spiritual omnipresence of the supreme being. To say that God was infinite and everywhere and then attribute the property of extension solely to matter was unwarranted and illogical.

In his *Immortality of the Soul* Henry More in 1662 went so far as to define spirit:

> I will define therefore *spirit* in general thus: a substance *penetrable and indiscerpible* [invisible]. The fitness of this definition will be better understood, if we divide *substance* in general into these first kinds, viz. *body* and *spirit*, and then define body as a *substance impenetrable and discerpible*. Whence the contrary kind is fitly defined as a *substance penetrable and indiscerpible*.

Penetrable and invisible spirit endowed space with an innate reality independent of the presence of impenetrable and visible matter. The fusion of the atomist and medieval universes, so important in the development of the Newtonian mechanistic universe of atoms and universal gravity, is usually ignored, because nowadays it clashes with our views concerning the unreality of spirit in the modern physical universe. Scientists feel embarrassed on learning that the noble Newton occasionally dabbled in mysticism, and they overlook the fact that if Newton had lacked a mystical outlook there could not have been a Newtonian universe. We always write history in such a way that its events are rational to us rather than the actors.

The divestiture of space of spirit was the work not of the Newtonians but of the deists who followed in the eighteenth century.

We do not know who was the first to identify gravity at the Earth's surface with the force reaching out and pulling on the Moon, and more generally, with the force reaching out from the Sun and pulling on the planets. (Perhaps Robert Hooke was the first.) An inner desire to reach the center of the Earth—a general appetition for wholeness—had to be transformed into a universal force issuing from each body and attracting all other bodies.

Without doubt, the spiritual emanations of the medieval universe, retained and refined by the Newtonians, inspired the flight of fancy that linked earthly appetition with universal gravity.

□　　□　　□

Robert Boyle, Edmund Halley, Robert Hooke, Isaac Newton, and Christopher Wren rank among the principal Newtonians. Consider Wren, at the age of 25, giving a lecture at Gresham College in 1657 and in the name of Seneca prophesying:

> A time would come when men should be able to stretch out their eyes, as snails do, and extend them fifty feet in length;

by which means they should be able to discover two thousand times as many stars as we can; and find the galaxy to be myriads of them; and every nebulous star appearing as if it were the firmament of some other world, at an incomprehensible distance, bury'd in the vast abyss of intermundious vacuum.

We remember Wren as an architect, for his design of St. Paul's Cathedral built after the Fire of London in 1666, and not for his mathematics and imaginative contributions to science.

Consider Halley, famed for his discovery of the cyclic return of Halley's Comet that we shall next see in 1986. Three decades after the death of Galileo the pace of scientific development was breathtaking; Halley had set up an observatory in the Southern Hemisphere, discovered the movement of stars, detected for the first time a globular cluster of stars, and plotted the paths of comets.

Consider the indefatigable Hooke, a thorn in the side of Newton, who for a time was Curator of Experiments at the Royal Society. Even greater were his accomplishments; he performed numerous ingenious experiments, made many suggestions concerning practical devices, such as the design of steam engines, and pioneered the microscope (introducing the word cell into biology); he advanced an array of ideas for explaining a variety of phenomena, developed a theory of earthquakes, proposed a wave theory of light, and opened up a grand vision of universal gravity.

Bodies move in straight lines when free of applied forces, as shown by Descartes, said Hooke, and bodies in circular motion are subject to a centrifugal force, as is well known to all. The planets do not move in straight lines but in circular orbits about the Sun; hence the planets experience a centrifugal force urging them away from the Sun and must therefore be continually pulled back toward the Sun by the force that commonly is called gravity. In 1670 Hooke explained his system of the world:

> That all celestial bodies whatsoever have an attraction or gravitating power to their own centres, whereby they attract not only their own parts, and keep them from flying from them, as we may observe the Earth to do, but that they do also attract all other celestial bodies that are within the sphere of their activity, and consequently that not only the Sun and Moon have an influence upon the body and motion of the Earth, and the Earth upon them, but that Mercury, Venus, Mars, Jupiter, and Saturn also, by their attractive powers, have a considerable influence upon its motions, as in the same manner the corresponding attractive power of the Earth hath a considerable influence upon every one of their motions also.

Hooke's "System of the World," which would lead, he said, to "the true perfection of astronomy," is the first statement on record concerning universal long-range gravity. (It is, of course, possible that he was not the first to make the discovery.) By 1679 Hooke knew from Christiaan Huygen's mathematical treatment of centrifugal force and Kepler's third law of planetary motion that the Sun's gravity weakens as the square of the distance from the Sun. He lacked the mathematical ability and rigor of a disciplined mind to convert his descriptive account into a system of precise laws.

□ □ □

A new epidemic plague in 1665 caused Newton at the age of 23 (and 23 years after the death of Galileo) to leave Cambridge and spend 2 years on his mother's farm at Woolsthorpe in Lincolnshire. During this period of enforced seclusion he investigated the properties of light, invented calculus, developed other mathematical subjects, and came to the conclusion, he later claimed, that gravity reaches out and controls the motion of the Moon around the Earth and of the planets around the Sun. Years later, looking back on this period, Newton said, "In those years I was in the prime of my age for invention, and minded mathematics and philosophy more than at any time since."

Natural philosophers had brought together in a higgledy-piggledy fashion a host of thoughts and discoveries; it was the astounding genius of Newton reflecting deeply for many years that synthesized these thoughts and discoveries into an organized whole. He enunciated the laws of motion in clear and precise form, taking into account the reciprocal action of forces, and he gave the theory of universal gravity a secure theoretical foundation. His great work *Mathematical Principles of Natural Philosophy* (written in Latin and known as *Principia*) was published in 1687. From a few principles and definitions Newton developed various propositions, then explained mathematically the elliptical orbits of planets, the twice-daily tides on Earth due to the attraction of the Moon and Sun, the equatorial bulge of the Earth owing to its rotation, and so on, until it seemed that all terrestrial and celestial phenomena were governed by mathematical laws of motion in a universe where each part gravitationally influenced all other parts in a precise and deterministic manner.

A universe created by God in which the heavenly bodies, separated by abysses of empty space, glided serenely according to supernal laws revealed to the mind of man; a universe, said Newton, "whose body nature is, and God the soul."

"Whence is it that Nature does nothing in vain and whence arises all the order and beauty in the world," wrote Newton in his *Opticks*, and a few lines later, "is not infinite space the sensory of a Being incorporeal, living, intelligent, omnipresent?"

□ □ □

Atomism, a theory of bygone ages (developed by the "most celebrated philosophers of Greece and Phaenicia," said Newton), did not enter the mainstream of science until the seventeenth century. The supple-minded Newtonians in their spirit of compromise converted the atheistic theory into a theistic theory of the universe. The Cartesians refused to accept the atomic theory of matter because it required that atoms were separated from one another by voids of empty space, and voids, as the Cartesians knew, could not exist. But the Newtonians, armed with the idea of space existing naturally of its own accord by virture of pervading spirit, had no fear of voids and were keen on the atomic theory, which Boyle applied enthusiastically with great effect to the study of gases. In his *Opticks* Newton said:

> It seems probable to me, that God in the beginning formed matter in solid, massy, hard, impenetrable, moveable particles, of such sizes and figures, and with such properties, and in such proportion to space, as most conduced to the end for which he formed them . . . even so very hard, as never to wear or break in pieces; no ordinary power being able to divide what God himself made one in the first Creation.

He had much more to say on the atomicity of matter, and he even proposed that light is composed of small particles.

Analysis of the scientific method is rarely helpful when we try to understand those central and creative figures—Pythagoras, Anaxagoras, Newton, and Einstein—who shaped and directed the advance of science. There are, of course, freewheeling periods in history when science seems to consist of little more than puzzle solving, of reaping the benefits of a reigning set of shared preconceptions, which Thomas Kuhn has popularized as paradigms. Innovative minds exploit the paradigms, and diligent investigators explore their consequences. Then, lo! Along comes a person with an original style of thought who envisions a new scheme of things, transforms the preconceptions, and lays the foundations of yet another universe. How does this person wind an armature of coiled themes, soldering the connections in such an artful manner that it generates sparks and lights up a whole new world of knowledge? We do not know, and possibly never will, for the strategies of truly creative thought lie outside the bounds of defined methodologies.

☐   ☐   ☐

Richard Bentley, a young clergyman who later became one of the great classics scholar of the eighteenth century, in 1692 gave a series of lectures entitled *A Confutation of Atheism,* in which he aimed to show how the marvels of the Newtonian universe provided evidence of the existence of God. Before publishing the lectures he asked Newton for his comments, and in the ensuing correspondence, Newton remonstrated, "You sometimes speak of gravity as essential and inherent in matter. Pray do not ascribe that notion to me. For the cause of gravity is what I do not pretend to know and therefore would take more time to consider it." From his various remarks, and the famous "I feign no hypotheses" in the second edition of *Principia,* it is apparent that Newton was unwilling to regard gravity as purely physical in its nature. Gravity, though quantifiable, was immaterial and mysterious; its presence furnished evidence of God's influence at work in the universe, and Newton shared Bentley's belief that gravity gave proof of God's existence.

The new theory of gravity was remarkable, because it reinforced the idea that the universe is centerless and edgeless, and therefore uniform and infinite. In one of his letters to Bentley, Newton wrote:

> It seems to me, that if the matter of our sun and planets and all the matter of the universe were evenly scattered throughout all the heavens, and every particle had an innate gravity toward all the rest, and the whole space throughout which this matter was scattered was but finite, the matter on the outside of this space would, by its own gravity, tend towards all the matter on the inside, and by consequence, fall down into the middle of the whole space, and there compose one great spherical mass. But if the matter were evenly disposed throughout an infinite space, it could never convene into one mass; but some of it would convene into one mass and some into another, so as to make an infinite number of great masses, scattered at great distances from one another throughout all that infinite space. And thus might the sun and fixed stars be formed. . . .

As Newton said, if the universe is of limited extent and has an edge and also a center of some sort somewhere, the attraction between the various parts would cause them "to fall down into the middle . . . and there compose one great spherical mass." But in a universe of unlimited extent, without center or edge, there is no preferred direction in which each part can be pulled. In the second edition of the *Principia* we find, "the fixed stars, being equally spread out in all parts of the heavens, cancel their mutual pulls by opposite attractions."

Bentley's task of confuting the atheists seemed not too difficult, either to him or Newton; mysterious gravity in a world of inert matter was undeniable proof of the handiwork of God. From the omnipresence of God in the medieval universe had come the infinity and uniformity of the Newtonian universe. Gravity, moreover, proved that the universe had no edge, was therefore infinite, and hence proved that God who had created the universe must be infinite also.

Thus the Newtonians paid back their debt. Gravity, derivative from the notion of pervasive theistic spirit, clearly demonstrated that the universe was necessarily infinite and uniform, and in turn demonstrated that the Creator, who could not be less than the created universe, was indeed infinite and everywhere. No other proof of the existence and nature of God has ever matched the elegance and self-consistency of that offered by the Newtonians.

7

The Mechanistic Universe

The telescope, microscope, thermometer, barometer, precision clock, air pump, and other inventions in the seventeenth century signaled the birth of the Age of Reason. Should the reader be unimpressed by these accomplishments in the "servile arts," why, then, the new age commenced with Descartes doubting all except the irreducible "I think, therefore I am." The Age of Reason—or the Enlightenment—perhaps began with the founding of the Royal Society in London and the Royal Academy of Sciences in Paris, or better still, with Isaac Newton taking refuge on his mother's farm and "voyaging through strange seas of thought alone."

The age of enlightened reason was ushered in by prophets proclaiming a new universe: "I feel engulfed in the infinite immensity of spaces whereof I know nothing and which know nothing of me, I am terrified. . . . The eternal silences of these infinite spaces alarms me," said Blaise Pascal; "Behold a universe so immense that I am lost in it. I no longer know where I am. I am just nothing at all. Our world is terrifying in its insignificance," said Bernard de Fontenelle.

The mechanistic universe of the eighteenth century more than fulfilled the promise of the prophets. Outfitted with laws uniting the Earth and the heavens, with self-running celestial systems distributed throughout endless space, with time ticking away regularly as in Huygens's precision pendulum clock, the mechanistic universe opened up the prospect of the human mind able at last to solve all the riddles of nature.

Lofty thoughts that once soared among the towers of the Eternal City descended to street level in the exhilerating new Earthly City. Otherworldly preoccupations transformed into worldly occupations. The world seemed bright and young, free of the dead hand of the Renaissance period with its conviction that all was senile and exhausted. Rejuvenated human sciences surged forward, led by "lapsed Christians" struggling against state and church for reforms that nowadays we take for granted as characteristic of Western society and Christianity. Law and justice, crime and punishment received fresh scrutiny in the light of reason; slavery practiced overseas by Europeans drew increasing condemnation; novel political credos inspired the framing of constitutions and bills of rights; lyrical and dramatic arts gave birth to essayists and novelists.

Also, deism supplanted theism among the enlightened.

Science inherits the Newtonian custom of referring to the reign of law and order. When scientists speak of the laws of nature, they have in mind those properties such as the laws of motion and the inverse-square law of gravity that when prescribed reveal regularity and harmony. To the Newtonians who were dyed-in-the-wool theists the world was God's temple, and the reign of law and order meant nothing less than governance by a supreme being. The Newtonians peppered their works with acknowledgments to the Almighty, the Supreme Wisdom, the Ruler who had created the universe, was manifest in its wonder and glory and continually at work in the execution of its laws and the maintenance of its order.

As the Age of Reason unfolded in the eighteenth century, the direct participation of a supreme being became less apparent. Laws of God denoting theistic superintendency transformed into Laws of Nature implicit in Nature itself; the theism of God's governance transformed into the deism of Nature's governance. God the First Cause, the Architect, the Author who had created the mechanistic universe was no longer employed as a maintenance mechanic. The universe of perfect law and sublime order was self-running and self-lubricating. The product was so good, as a current commercial says of a certain washing machine, it required no maintenance. From the self-running mechanistic universe, God withdrew into a vague background of abstract being and remained there as the indispensible architect of it all. "If God did not exist," said Voltaire the deist, "it would be necessary to invent him."

"Cubic space-division," by M. C. Escher. This picture can be viewed as a grim representation of the infinite mechanistic universe braced by rigid laws of nature. (© BEELDRECHT, Amsterdam/V.A.G.A., New York, Collection Haags Gemeentemusem—The Hague, 1981.)

By emphasizing Nature and Nature's laws, the deists avoided direct reference to God and God's laws. The relation of human beings to Nature usurped the relation of human beings to God. The changeover from theism to deism opened up for exploration territory previously fenced off as holy ground; by searching for human-nature laws in the mechanistic universe, renascent human sciences (sociology, anthropology, psychology, economics) strived to emulate the successes of the natural sciences.

The Age of Reason in the eighteenth century brimmed with bright hopes, bubbled with utopian dreams, overflowed with youthful ebullience. Despite its ups and downs, its turbulent radicals such as Thomas Paine demanding "life, liberty, and the pursuit of happiness," it was a period of moderate political equilibrium, of surging economic and industrial growth, of high finance and booming overseas trade. The harnessing of natural science to industry, and the development of powerful steam engines, gave impetus to an industrial revolution that had its roots in the Middle Ages. Even the churches ceased to harry and torment heretics, and the last witches met their doom in Western Europe in the early decades of the century. In the new universe with its new god such horrors were inhuman and ungodly.

Instead of being witch-crazy the Europeans became project-crazy. Everyone, it seems, had a pet scheme: business projects, get-rich-quick projects, prison-reform projects, educate-the-poor projects, pave-the-roads projects, welfare projects, emancipation projects, get-rid-of-the-aristocrats projects, make-everyone-an-aristocrat projects, projects to establish overseas colonies, and hosts of others of every kind abounded, all championed with enthusiasm and optimism.

In the air was the heady realization that the ancient world had been overtaken in every field of human endeavor. And indeed it was true. A many-sided civilization had emerged of altered mentality, of numerous minds striving individually and collectively, equipped with a formidable universe potentially more powerful and magnificent than any seen before.

It was a cuckoo universe, soon to attain such compelling power that it would enforce worldwide adoption, throwing out and replacing the indigenous belief-systems of non-European nations.

□   □   □

The Renaissance had only the dream, the hope of glimpsing and rivaling the wisdom and gracious living of classical antiquity. It was bereft of thoughts on the possibility of progress. Through the eighteenth century swept the notion of progress like a wind sweeping away the cobwebs and dust of ages. Society hummed with purposeful activity, and everywhere ran an awareness that things were going places.

Deistic historians traced the ascent of man, seeking to understand where human beings had come from in order to plot where they were going. To this end they reconstructed the whole of history and outfitted their virgin universe with a past as new as a bride's trousseau. Not just the old past with a few amendments and the latest chapter added, but a new past, in which the shuttle of the human story wove a fabric of novel design. A new universe, the deists discovered, needed a new history.

In the language of the revised history, religion translated into superstition, evil into ignorance, redemption into reason, divine grace into human virtue, God into Nature, Providence into Progress, and last but not least, Judgment into Posterity. In the new history the Garden of Eden was symbolic of a golden age of Noble Savages, the Fall symbolic of the rise of organized religion and the tyranny of priests, and Judgment symbolic of the esteem of Posterity for prestigious works. The Elected— the Europeans—led out of Exile by the Goddess of Wisdom, could now look forward to a Promised Land, overflowing with sanity and happiness, filled with the prospect of the perfectability of man. Thomas Jefferson and Benjamin Franklin shared these views and thought that if only human beings could rise above their religious preoccupation with the sinfulness of human nature and be free of their obsession with an afterlife, then all obstacles would vanish in the path of social progress.

The ancients regarded time as cyclic, with the endless return of golden ages alternating with dark ages. What had happened yesterday and yesterday and yesterday will happen tomorrow and tomorrow and tomorrow. The repetition of the Wheel of Time was nonprogressive. But the Persian idea of time, adapted to the medieval universe, consisted of a single cycle that began yesterday with Creation and will end tomorrow with Judgment. The deists were the first to jump off the treadmill of cyclic time. The concept of progress—progressive improvement and continual development—changed time and made it open-ended.

According to the new universal history, God created the universe in the beginning, and thereafter time had ticked away as in a well-oiled clock. The Apocalypse, the Coming, and the End were out of sight, for the celestial machinery could never wear out. A bright future of unlimited progress stretched ahead in an unbroken expanse of time, free of Doomsday and Judgment, in which all the accomplishments thus far would fade into nothing compared with those to come.

Instead of Judgment and the award of treasure in Heaven, the deists believed in Posterity and the award of treasured memory on Earth.

[105]

Judgment by posterity, which I shall call the *principle of pharaonic immortality*, has become one of the main driving forces in the modern world.

"The king whose achievements are talked about does not die," recorded Sesostris I, and the inscribed annals of Thutmose III were placed in his sanctuary "that he might be given life for ever." The deists, like the Egyptian pharaohs of the second millennium B.C., believed that immortality lies not in another world but in the memory of generations to come.

Belief in pharaonic immortality has gained widespread acceptance in modern times. The heavenly rewards of the Eternal City have become myths; in their place honors, prizes, and Emmy awards abound in the Earthly City, nourishing the need for fulfillment and serving as insignia for the attention of posterity. When distinguished persons die, glowing obituaries and biographies are written, memorials erected, and commemorative prizes instituted. Thus is their memory preserved, and they have life ever after.

It was once God who saw everything and rewarded good works. Now society judges, posterity rewards, and publicity not prayer is what really matters. We no longer hold in our conscious minds the steadfast belief that Someone is watching, recording our motives and deeds, who one day will judge us impartially and independently of what other people think. The internal watchdog who maintained the highest standards and could not be deceived has gone. Instead, each person strives for recognition by society and endeavors to build a personal pyramid that will enshrine and preserve his or her memory. Publicity is all that counts, and the worst thing that can happen is to be ignored. Why do writers fill libraries with books, scientists seek to disturb the universe, architects reshape the landscape, artists, historians, politicians and the rest try to make their mark in memorable works, and criminals find gratification when their deeds are trumpeted around the nation? The mainspring of this dynamism is the desire for pharaonic immortality.

The Age of Reason faltered with the rise of a romantic movement in revolt against savants and their mechanistic blueprints, with Blake's mysticism, and with Wordsworth's despairing cry "We murder to dissect." It certainly had reached low ebb in the early nineteenth century when Thomas Arnold, historian and headmaster of Rugby, wrote in a letter, "Rather than have Physical Science as the principal thing in my son's mind, I would have him think that the Sun went round the Earth, and the Stars were mere spangles in a bright blue firmament." The enlightened honeymoon had ended with the French Revolution, and in the wars that inundated Europe, human beings were as benighted as ever.

☐   ☐   ☐

Across the vault of heaven stretches the wraithlike arch of the Milky Way—the via lactea—formed by the numerous stars and luminous gas clouds of our Galaxy. Thomas Wright of Durham believed that the Milky Way offered ample reason for glorifying the works of God.

Wright was a gadabout youth of 16 years when Newton died in 1727 at the age of 85. After settling down in marriage as a surveyor and a teacher of mathematics to "noble ladies," he turned his attention to the spectacle of the heavens. At first he agreed with Newton and thought the stars were "promiscuously distributed through the mundane space." Later he realized that the observed stars are not randomly scattered but appear to be arranged "in some regular order." He published his thoughts in 1750 in a book entitled, *An Original Theory or New Hypothesis of the Heavens, Founded on the Laws of Nature, and Solving by Mathematical Principles the General Phenomena of the Visible Creation; Particularly the Via Lactea.*

Wright proposed two models of the Galaxy, and in the one of interest to us he arranged the stars in a disklike system rotating about a cosmic center. The Milky Way was the disk of stars seen from our position inside the disk. Being a diehard theist, he viewed the universe as an arena of theistic activity and proposed a Neoplatonic galactic center endowed with supernatural power. "At this center of creation," he wrote, "I would willingly introduce a primitive fountain, perpetually overflowing with divine grace, from whence all the laws of nature have their origin." A deist at this stage might have thrown the book aside in disgust and missed the most daring of Wright's conjectures. Wright went on to suggest that the fuzzy and faint nebulae of the night sky are perhaps other creations or "abodes of the blessed," similar to our Milky Way, but far away.

In the agile mind of Wright, Newton's universe of scattered stars had transformed into an endless vista of "abodes of the blessed," each a distant and gigantic system of stars like our Milky Way.

Immanuel Kant in the university town of Königsberg read a review of Wright's work. Four years later in 1755 he published his own book having the equally long title, *A Universal History and Theory of the Heavens; An Essay on the Construction and Mechanical Origin of the Whole Universe, Treated According to Newton's Principles.* In this work Kant presented the most stupendous universal picture ever conceived.

According to Kant's version of Genesis, in the beginning was chaos, as proposed by "the ancient philosophers," and like those philosophers he assumed that the "first state of nature consisted of a universal diffusion of primitive matter, or of atoms of matter, as those philosophers have called them." Out of the vortical motions of chaos, under the influence of gravity, came stars that congregated to form the Milky Way. Kant then drew on Wright's suggestion. The distant nebulae seen in the night sky as small elliptical patches of fuzzy light were whirlpool milky ways at great distances, each similar to our Milky Way. He went further: not only were the stars clustered into milky ways (now called galaxies), each held together by its own gravity, but the milky ways were themselves clustered together to form much larger systems (or clusters of galaxies), each also held together by its own gravity. The clusters of milky ways, continued Kant, were clustered to form much larger systems that in their turn were clustered to form yet vaster systems, and so on, in an endless progression of systems of increasing size, filling infinite space.

Kant quoted Pope,

> Look around the world; behold the chain of Love
> Combining all below and all above

and saw in his hierarchical universe a natural extension of the great chain of being. "The theory we have expounded opens up to us a view into the infinite field of creation, and furnishes an idea of the work of God which is in accordance with the infinity of the great Builder." Unlike Wright, Kant was a deist and believed the created universe so perfect that further theistic intervention was unnecessary: "God has put a secret art into the forces of nature so as to enable it to fashion itself out of chaos into a perfect world system."

William Herschel, born in Germany, lived in England from 1757 onward; aided by his sister Caroline he became the foremost astronomer in the Age of Reason. Both abandoned musical careers because of their consuming interest in astronomy, and both devoted their lives to constructing telescopes and observing the heavens. Discovery of the planet Uranus brought fame to William. He was fond of pointing out that astronomy has much in common with botany. "The heavens are seen to resemble a luxuriant garden, which contains the greatest variety of productions." Stars evolve and have life histories, and at a glance we see them in their various stages of development. "Is it not the same thing," he wrote in *The Construction of the Heavens*, "whether we live successively to witness the germination, blossoming, foliage, fecundity, withering, and corruption of a plant, or whether a vast number of specimens selected from every stage through which the plant passes in the course of its existence be brought at once to our view?" Because we cannot wait for an acorn to evolve into an oak tree, we can at least observe oaks in various stages of growth and piece together the life history of a typical oak tree. Similarly, astronomers cannot wait for a star to evolve and must piece together the life history of a typical star from the display of many stars in various stages of evolution.

The Herschels observed and cataloged numerous stars and nebulae and were the founders of modern astronomy. William, a true son of the

Enlightenment, not surprisingly had many simplistic beliefs; he thought it obvious that the Moon was inhabited and that beneath the bright atmosphere of the Sun was possibly a cool surface also populated with living beings.

When Napoleon Bonaparte became first consul of France in 1799, he appointed Pierre de Laplace, a mathematician, as minister of the interior, then fired him 6 weeks later for creating a bureaucratic nightmare by attempting to introduce "the spirit of infinitesimals into administration." When Laplace presented to Napoleon a copy of his great work *Celestial Mechanics*, Napoleon said on this occasion, "You have written this huge work on the heavens without once mentioning the Author of the universe." To which Laplace replied, "Sire, I had no need of that hypothesis." In the sciences, henceforth, God was relegated to the role of designing the laws and molding the atomic parts, but not required to appear in person.

□ □ □

In the seventeenth century the Newtonians finally broke the bonds of medieval space, and in the boundless expanse of a new universe, human beings lost their privileged central location. World pictures were severely jolted. But sensible persons quickly realized that little had actually been lost, for human beings still figured prominently in the cosmic design and remained the most conspicuous members of the Great Chain of Being that linked them directly to the throne of God. Men and women still retained their central location in the much more important biological-spiritual universe.

In the nineteenth century the savants of the mechanistic universe finally broke the bonds of medieval time, and the Beginning receded into the mists of a remote past. Genesis was controverted and the Great Chain crashed down. Organic life, enmeshed in the cosmic gearwheels, became an integral part of the mechanistic synthesis of dead matter. God's temple of the Newtonian universe collapsed, and all that had seemed of highest value lay crushed in the ruins, empty of cosmic meaning.

World pictures became neurotic, even psychotic, and the consequences are apparent in the distress and pathological disorders of the twentieth century. Some psychologists allude to the possibility that psychosis is a natural healing process; this is certainly true in the case of world pictures unhinged by a transformation of universes. A transitional period releases deranged individuals from an unbearable reality that is being forced on them by the rise of a new and frightening universe. They rally to extremist groups, form iconoclastic movements against-this and down-with-that, revert to antiquarian religions, flock to political creeds that purport to give cosmic significance to life, grieve in counterculture communities, or retreat into autistic worlds of secret knowledge.

It is hopeless trying to understand the nineteenth century, with its fulminations from pulpit and platform, without realizing that numerous persons were struggling to save their imperiled world pictures that gave meaning and purpose to life on Earth; nor can we hope to understand the furor of the scenes enacted in the present century unless we realize that many societies were struggling, and still are, to find new beliefs, often with dismal and tragic results.

In the nineteenth century the entire mythic universe was on trial; at stake was the veracity of biblical records indicating that creation had occurred a few thousand years ago. The Mosaic chronology of scriptural records (derivative but erring from Babylonian chronology) sustained the deep-rooted belief that human beings were of paramount importance in the cosmic scheme. From biblical records Dante estimated that the creation of Adam occurred in 5198 B.C.; Kepler estimated that the creation of the world happened in 3877 B.C.; James Ussher, an Irish bishop, fixed the date of creation at 4004 B.C.; and Newton, in his *Chronology of Ancient Kingdoms Amended,* set the date at 3988 B.C.

Astronomy had regrettably been mechanized; little was lost, however, for the heavens still proclaimed the glory of the Lord and conformed to providential law. But geology probing into Earthly history was quite another matter. Here was a subject, outside the purview of natural law, in which miracles once had free play. At the gates of geology, said Thomas Huxley, "stood the thorny barrier with its comminatory notice— No Thoroughfare. By Order, Moses." In the thickets of earthly history lay proof of the veracity of biblical records, and geologists and natural historians arguing against the brevity of life on Earth were heretics, if not downright atheists, seeking to disprove the truth of holy writ.

We must backtrack a little. Georges-Luis Leclerc de Buffon, keeper of the Jardin du Roi in the mid-eighteenth century, proposed that a large comet had struck the Sun a glancing blow and that the ejected matter then condensed to form the planets of the Solar System. He estimated the Earth had taken 100,000 years to cool to its present temperature. To reconcile this calculation with Genesis, he supposed that each of the 6 days of creation was actually a period of long duration, and "day" needed reinterpretation in the light of new knowledge. In his masterpiece of 1778, *The Epochs of Nature,* Buffon rolled back biblical time to a remote beginning and said that natural history is revealed in the archives of nature and must be regarded as a science on the same footing as astronomy. His seminal ideas concerning the antiquity of Earth and the evolution of prehistoric life, and his suggestions that coal deposits are the remains of prehistoric life, and the Great Chain of Being a web of interconnecting links like chain mail, provoked more outrage than he bargained for.

Denis Diderot, a French contemporary encyclopedist and a notorious free-thinking philosopher, argued that the work of Kant clearly showed that the age of the universe is not just hundreds of thousands of years but more probably "hundreds of millions of years." Nowadays the age of the Solar System is reckoned to be in the region of 5 billion years, and the age of the universe somewhere between 10 and 20 billion years.

By the beginning of the nineteenth century it was impossible for natural historians to brush aside the accumulation of evidence from the study of fossils and rock strata. Compromise doctrines capable of accomodating the Mosaic chronology became the fashion. Georges Cuvier, a distinguished French naturalist and later chancellor of the University of Paris, argued that the Flood was a crucial event separating supernatural and natural history. Human beings were created just before the onset of natural history, and the soul-less lifeforms of the fossil record lived in the antediluvian periods. Further elaborations soon became necessary. The globe had apparently been periodically visited by many catastrophes, such as life-destroying deluges, and newly created life had arisen in more advanced forms after each visitation had devastated the globe. Thus the control of natural and supernatural laws alternated, and life was created anew in episodic acts of special creation.

The Scottish physician James Hutton had little patience with Mosaic chronology and proposed a uniformitarian doctrine. The geological record reveals, he said, continuous erosion and sedimentation acting over vast periods of time, so vast that there "is no vestige of a beginning—no propect of an end." Declaring, "no powers are to be employed that are not natural to the globe, no actions to be admitted except those of which we know the principle," he laid in 1785 the foundations of modern geology.

Hutton's deistic picture, untrammeled by catastrophic acts of theistic intervention, was later adopted by the great Scottish geologist Charles Lyell in 1830 as the theme of his *Principles of Geology*. The upthrust and erosion of mountains, the sculpturing of landscapes, and the shaping of continents are the result of steady and natural processes acting over interminable ages, wrote Lyell. "Thus, although we are mere sojourners on the surface of the planet, chained to a point in space, enduring for a moment in time, the human mind is not only enabled to number worlds beyond the unassisted ken of mortal eye, but to trace the events of indefinite ages before the creation of our race." In like manner, he continued:

We aspire in vain to assign the works of creation in *space*, whether we examine the starry heavens, or the world of minute animalcules which is revealed to us by the microscope. We are prepared therefore to find that in *time* also the confines of the universe lie beyond the reach of mortal ken. But in whatever direction we pursue our researches, whether in time or space, we discover everywhere the clear proof of a Creative Intelligence, and of His foresight, wisdom and power.

On the one hand, the catastrophists believed in periodic cataclysms followed by acts of special creation, of which the last act occurred only a few thousand years ago; on the other hand, the uniformitarians believed in a single created state of long ago, and natural laws have since held uninterrupted sway. Catastrophe versus uniformity sounds nowadays much like big bang versus steady state, but the controversy that raged in the early decades of the nineteenth century was far more heated than the debate between big-bangers and steady-staters in recent decades. The catastrophists lost in the last century but have won in this century.

□　　□　　□

The notion of evolution was in the air, affecting the climate of opinion. Most members of the public accepted social evolution as synonymous with progress. One had only to compare the lifestyles of civilized and uncivilized people to see that social evolution obviously occurred. Across the gap separating European and primitive cultures stretched a chain of progressive social evolution. It was not social but organic evolution that caused all the trouble.

The century of evolution—the nineteenth century—began with Jean-Baptiste de Lamarck, a French naturalist who popularized the word biology. He resuscitated the old idea that organic life evolves from rudimentary beginnings and advances as it adapts to the exigencies of the environment. The cardinal idea of Lamarckian evolution was that creatures evolve organically in response to their needs and desires, and evolution is, therefore, self-directing. Common sense urges us to believe that skills and aptitudes acquired by parents are inherited by their offspring. We feel that progress of this kind is the most valuable thing that can be handed on to our descendants. Such was the thrust of Lamarck's argument. But common sense is wrong, and so was Lamarck, and in matters concerning evolution common sense often misleads us. If you believe that by your study and athletics your children will be studious and athletic, you are wrong; at most they will benefit by your example.

Robert Chambers, one of two brothers who founded the *Chambers'*

Encyclopedia, was a persuasive writer who popularized the Larmarckian theme. He performed, in Darwin's words, valuable service "in removing prejudices, and in thus preparing the ground for the reception of analogous views." Robert Chambers' widely read book *Vestiges of the Natural History of Creation* was as sensational in its day as Darwin's *Origin of Species* 15 years later. Competition and the struggle for life dominated the whole period of prehuman history, explained Chambers, and "the adaptation of all plants and animals to their respective spheres of existence was as perfect in those earlier ages as it is still." The struggle to survive and the ensuing adaptation by plants and animals to the vicissitudes of the environment was in full accord with natural law. Theistic intervention was quite out of court:

> We have seen powerful evidence that the construction of this globe and its associates, and inferentially all other globes of space, was the result, not of any immediate or personal exertion on the part of the Deity, but of natural laws which are expressions of his will. What is to hinder our supposing that the organic creation is also a result of natural laws, which are in like manner an expression of his will?

The audience addressed by Chambers was more enlightened than we sometimes realize; at that stage, of course, human beings still retained their own special creation and in no way were "descended from monkeys."

At about this time Herbert Spencer developed a comprehensive scheme of evolution embracing biology and sociology. Like Chambers he advocated a principle of "law versus miracle" and argued that evolution is "not an accident but a necessity." He used the phrase "survival of the fittest" that has since been gratuitously affixed to the Darwinian theory of natural selection. As with others of his day, Spencer had difficulty distinguishing between evolution (denoting natural change) and progress (denoting improvement as judged by value concepts). Evolution, said Spencer, moves "from an indefinite, incoherent homogeneity to a definite, coherent heterogeneity" and is the development of systems of greater differentiation possessing greater integration.

We may intepret Spencer's theory rather freely as follows. Let the degree of integration of a system—either an organism or a society—be represented by the symbol X and its degree of differentiation by the symbol Y. The product XY increases with progress and is a measure of the excellence of the system. When X and Y are both large, the system has simultaneously a high degree of integration (X is large) and a high degree of differentiation (Y is large); in other words, the system is generalized and yet specialized. Thus in a society if X and Y are both large, we have social order and individual freedom; if X is small and Y large, then disorder and anarchy; and if X is large and Y small, then order and totalitarianism (note that fascism, communism, and socialism imply large X but not necessarily large Y).

In Western society we tend to stress the importance of organization and standardization, and X is undoubtedly now larger than in previous centuries. But individual freedom has not increased to the same extent and may actually be decreasing. The product XY may not be much larger than before and our progress, therefore, less than we suppose. Uniformity can easily be attained, as in time of war, when freedom to be different is sacrificed for the benefit of greater integration. But a harmonious society in which individuals have freedom to be different is much more difficult to attain. Social progress requires that integration and differentiation increase, or at least their product increases, and not just one at the expense of the other. Spencer's argument on integration and differentiation as the hallmarks of progress is intriguing, especially when in a reflective mood we gaze at the scurrying of an ant heap or at the hustle and bustle of a city and wonder how large is X and how small Y; so far it has not figured prominently in the study of organic evolution.

Spencer's equation ($Z = XY$, where Z is a measure of progress) is of interest in ethics. The tendency in modern society to transfer ethical responsibility from the individual to the state undoubtedly increases X, but regrettably decreases Y, and the value of Z may consequently not be as great as we often suppose. Greater emphasis on duties rather than rights might create a better balance and help to increase the value of Z.

☐ ☐ ☐

The evolutionary theory of natural selection was independently proposed in 1858 by Alfred Wallace and Charles Darwin. A year later, in his great work *On the Origin of Species*, Darwin presented his thoughts and investigations of many years. It suffices here to say that the theory of natural selection was inspired by four streams of thought.

The first is that organic life evolves naturally and is not subject to supernatural law. The second concerns the antiquity of the Earth; on this subject Darwin was influenced by Lyell and wrote, "He who can read Sir Charles Lyell's great work on the *Principles of Geology*, which the future historian will recognize as having produced a revolution in natural science, and yet does not admit how vast have been the past periods of time, may at once close this volume." The third, familiar to breeders, is the knowledge that members of a species are not all exactly alike but differ in various slight ways. The fourth is the realization that the growth of populations is checked by environmental limitations; this last stream of thought had its source in the Reverend Thomas Malthus's *Essay on the Principle of Population* written in 1798. It is a strange fact, observed Malthus, that the human population is held in check by war, disease, and premature death and is thus prevented from overburdening the available natural resources of society and the land.

From the confluence of these four streams came the mainstream of natural selection. Given natural laws and sufficient time, individual differences favoring survival and reproduction are shared increasingly among the members of an interbreeding population, and the species evolves. Though producing many strange twists and turns, resulting sometimes in bizarre lifeforms, the natural selection of individual variations by differential reproduction is as inexorable as any other law of nature.

Darwin did not understand the cause of individual variations within a species. We now know that the genetic coding in twin-stranded molecules of nucleotides determines organic structure; mutations in the coding are responsible for individual variations and are the consequences of natural rearrangements inevitable in complex molecular systems.

Through the genetic coding in the cells of living creatures the past reaches out and confronts the present. Natural selection is dynamic, because the lifeforms selected by past conditions now exist in present and different conditions. It is a game of tag in which the past never catches up. The naturally selected become the selected unnatural; the fittest survive and become unfit and do not survive.

Natural selection is as blind as the law of gravity and does not guarantee progress of any desired kind whatever. In fact, evolution has nothing to do with progress. Each step is dictated by what breeds the most, and whole species are blithely discarded that later, in an altered environment, would have been superior in fitness to those surviving.

Perhaps the Neanderthals were more intelligent than our Cro-Magnon ancestors, yet failed to survive because of inferiority in aggression. In that case their more intelligent nature and peaceful character would nowadays, in an age of nuclear weapons, count as assets in the survival game. Nothing in natural selection safeguards human beings from vanishing in the future or evolving into creatures of less intelligence.

In some ways Lamarck was right. Human beings do have control over their evolution, but not quite in the way he thought. The environment plays the tune, life dances accordingly, and humans fiddle with the environment. The natural environment is fast vanishing and being replaced with a dense matrix of human beings and their artifacts. The ransacked biosphere, now in a precarious state and devastated for the odd purpose of maximizing the number of human beings, is fast becoming derelict and catastrophically unstable. Modern Lamarckism, it seems, is as blind as natural selection; perhaps worse, for natural selection at least produced the human species, whereas Lamarckism at present is only capable of presiding over its demise.

☐   ☐   ☐

Evolution means "unrolling" or "appearance in orderly succession of a long train of events." The Sun evolves. Hydrogen in its deep interior slowly burns into helium, and in roughly 5 billion years the Sun will evolve into a red giant, then into a white dwarf star. Evolution in astronomy applies both to slow secular change of equilibrium configurations and to rapid episodic transformations, as in novae and supernovae. Stellar evolution does not imply "progress" of any kind, and no astronomer would dream of such an implication. Evolution, as used in astronomy, has a simple and straightforward meaning consistent with a mechanistic treatment in the physical universe.

In the biological sciences the word evolution is saturated with value concepts. The notion of social progress in the eighteenth century carried over into the notion of organic evolution in the nineteenth century. Progress became evolution, and evolution retained much of the meaning of progress. Progress means improvement, a change to better things, from lower to higher levels, and it involves value judgments that are empty of physical content. The notion that evolution generally is progressive is deeply embedded in the language of the biological sciences: things that evolve are things that advance and improve.

Owing to the aura of progress investing the notion of evolution, we use fittest, advantageous, and other such terms that are saturated with value concepts. When we try to justify our value concepts we find ourselves trapped in circular argumentation. Individuals surviving are the fittest, but what are the fittest? Obviously, those that survive. Individuals having advantageous variations reproduce and flourish, and what are advantageous variations? Obviously, those that reproduce and flourish. Whenever a value judgment trespasses into the physical universe, it chases its tail.

The seductive word evolution, haloed in the life sciences with the mystique of progress, has no place in the mechanistic processes of the physical universe. Biological evolution either takes place or does not take place in the physical universe; there can be no fudging with a halfway world that is neither one thing nor the other, not if we wish to be rational in the universe of our society. As I see it, the law of natural selection is a physical law and must be treated as such. My modest proposal is that evolution should be used only by astronomers, who have retained its proper meaning; natural historians would confuse us less if they stuck to natural selection and used safe words such as change and alteration.

The Origin of Species closes with the sentence, "There is grandeur in this view of life, with its several powers, having been originally breathed into a few forms or into one; and that, while this planet has gone cycling on according to the fixed laws of gravity, from so simple a beginning endless forms most beautiful and most wonderful have been, and are being, evolved." The clockwork universe of natural laws, extended endlessly in space and to the limits of time, now embraced all animate as well as inanimate things.

It was the grandeur of a mechanistic universe, awesome and awful, that banished in its chilly light the last shadows of the mythic universe, a universe in which shivering human beings stood and reassured one another of its glory and wonder. Edwin Burtt in *The Metaphysical Foundations of Modern Physics* writes,

> The world that people thought themselves living in—a world rich in color and sound, redolent with fragrance, filled with gladness, love and beauty, speaking everywhere of purposive harmony and creative ideals—was crowded now into minute corners in the brains of scattered organic beings. The really important world outside was a world hard, cold, colorless, silent, and dead; a world of quantity, a world of mathematically computable motions in mechanical regularity. The world of qualitites as immediately perceived by men became just a curious and quite minor effect of that infinite machine beyond.

At last we come to the twentieth century. Adrift like shipwrecked mariners, in a vast and meaningless mechanistic universe, we are found clinging for life to the cosmic wreckage of ancient universes.

Part 2

The
Heart
Divine

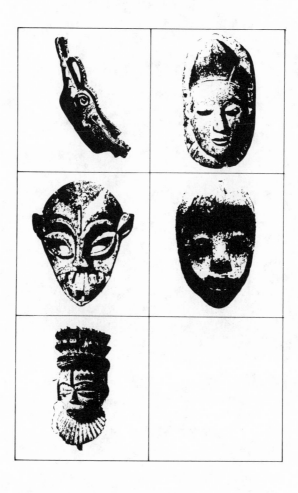

8

Dance of the Atoms and Waves

In everyday life we deal with things of sensible size—such as plants and flowerpots—and to understand these ordinary objects we explore the worlds of the very small and the very large. We delve into molecules and atoms and reach out to the stars and galaxies. Thus we know that most atoms composing the Earth were made in stars that died long ago before the birth of the Sun.

This wide realm of nature, consisting of things stretching in size from atoms to galaxies, is ruled not by the gods of antiquity, but by the laws of motion and the push and pull of electrical and gravitational forces. Electrical forces dominate on the scale of molecules and atoms, accounting for much of the intricacy of the very small; gravitational forces dominate on the scale of stars and galaxies, accounting for much of the intricacy of the very large.

The great problems, deep at the foundations of the physical universe, no longer may be found in the wide realm of nature stretching from atoms to galaxies. The exploration of this luxuriant garden of phenomena is in the care of sciences such as chemistry, biochemistry, geophysics, and astrophysics. The great problems of the physical universe are now found in what might be called the outer realms of nature. When the scale of measurement decreases a hundred thousand times smaller than the size of atoms and increases a hundred thousand times larger than the size of galaxies, we quit the lush middle realm and enter the austere outer realms. Here we discover the truly baffling: in the subatomic realm of the extremely small lies the enigmatic diversity of strange forces; in the cosmic realm of the extremely large lies the enigmatic unity of the physical universe.

In this chapter we start by looking at the atomic and subatomic realm.

☐   ☐   ☐

Pierre Gassendi of the seventeenth century, a French professor of philosophy and mathematics who believed that happiness consists of the harmony of body and soul, sought to revive the dormant atomic theory. He stripped from the atheistically tainted atomic theory of the ancient world its vehement negation of the gods. The Newtonians adopted the revived atomic theory and wove it into their theistically created mechanistic universe. After 2000 years the atomic theory had at last become respectable.

René Descartes and the Cartesians who followed in his footsteps would have nothing to do with the atomic theory; they insisted that matter was necessarily infinitely divisible and atoms could not exist. But Robert Boyle, the alchemist of a new age, showed how the idea of atoms explained the properties of gases; the final blessing was given by Newton, who said, there are "agents in nature able to make the particles of bodies stick together by very strong attractions. And it is the business of experimental philosophy to find them out." Physicists of the twentieth century follow Newton's advice and devote considerable time and effort to the business of finding out these very strong attractions in the atomic and subatomic world.

The atomic theory entered chemistry in the early nineteenth century. John Dalton, who investigated color blindness and was himself color blind, popularized the Greek word *atom*, meaning uncuttable. By supposing that atoms of different elements have different weights, Dalton showed how the elements combine in fixed proportions to produce chemical compounds. He was the first to make the atomic theory quantitative. The discovery of the electron in 1897 by Joseph Thomson and of the atomic nucleus in 1911 by Ernest Rutherford launched the modern era of atomic physics. Niels Bohr in 1913 constructed a model of the atom in which the electrons move in orbits about the nucleus; a few years later Louis de Broglie, Erwin Schrödinger, Werner Heisenberg, and other eminent physicists laid the foundations of the quantum mechanical model of the atom. The quantum world of the atom has transformed our view of the physical universe.

An atom consists of a small positively charged nucleus surrounded by a cloud of negatively charged electrons. Many persons may vaguely recall hearing such a statement while dozing in the classroom. Probably the teacher was not a poet. "When it comes to atoms," said Niels Bohr, "the language that must be used is the language of poetry." Students wake up when told about cells, cytoplasm, organelles, chromosomes, and the double helix, for here are things of human significance. They hear music in this litany of life-science terminology, but not in the jargon of electrons, protons, neutrons, and other lifeless creatures of the quantum world. It is a pity, for without atomic particles there could be no organic life, and in the impalpable and seemingly inconsequential entities of the quantum world, one finds the true music and magic of nature.

We have taken the first big step: the atom consists of a central heavyweight nucleus surrounded by a cloud of lightweight electrons—a step accompanied by many misconceptions. Once it was the custom to imagine the atom as a miniature solar system with electrons encircling the nucleus like planets orbiting a sun. This idea still persists in popular literature. But electrons do not move in clear-cut orbits like revolving celestial bodies. They dance, and the atom is a ballroom. The electrons perform stately waltzes, weave curvaceous tangos, jitter in spasmodic quicksteps, and rock to frenetic rhythms. They are waves dancing to a choreography composed differently for each kind of atom.

The old idea that subatomic particles are similar to billiard balls is out. Piet Hein in *Atomyriades* miscued when he wrote the lines:

> Nature, it seems is the popular name
> for milliards and milliards and milliards
> of particles playing their infinite game
> of billiards and billiards and billiards.

A particle is not like a billiard ball. It is a vibrant creature of a little world made cunningly.

The electron waves spreading out and interweaving wherever possible account for the structure of atoms. They lace together arrays of atoms into molecular tapestries and create the rich and varied patterns of our world of plants and flowerpots.

How can a tiny electron behave like a widespread wave? We must face the fact, as much a fact as any we know, that all subatomic particles, not only electrons, have a remarkable dual nature. At one moment a particle is like the ripples on the surface of a pond, and at the next moment like a cherrystone in the palm of the hand. A particle is both wave-like and corpuscular, and its dual nature is as perplexing as the duality of mind and matter. The strange quantum world contains nothing resembling our world of commonplace experience, and we must not try to comprehend things in the vulgar fashion. In *The Character of Physical Laws,* Richard Feynman remarks, "I think I can safely say that nobody understands quantum mechanics." He means that nobody understands with ordinary common sense. The new territory is bizarre; tourists marvel, and physicists take up residence.

Mostly, an electron is wavelike and widespread over regions of space. These regions, though sometimes large, are usually small as in atoms. An electron or any other subatomic particle fills all accessible space with its wavelike behavior and rarely acts as a discrete entity.

The electron spreads everywhere as a wave, and when we succeed in observing it, something odd happens: it collapses and becomes a sort of corpuscular creature active within a small region, and we can then say, Ah, now I know roughly where it is!

Consider the following weird situation. An electron is shot at a target. It travels not like a corpuscle but as a wave and has all the remarkable properties of waves. It strikes the target somewhere not as a wave but as a corpuscle and emits perhaps a scintilla of light betraying where it lands. We never know exactly the spot at which the electron will strike, and we can only estimate from its wavelike behavior the chance—or probability—of its appearance in a corpuscular form. The probability of where it will appear is proportional to the square of the amplitude of the wave. Where the wave is strongest (and does not interfere with itself) is where the electron has the best chance of being observed.

There is nothing chancy about the waves themselves. They are fully predictable, and in their complex ways they evolve from state to state and travel from place to place in a manner fully determined by the equations of quantum mechanics. But only from their amplitude can we estimate the probability of where and in what way the waves will collapse and become an observed event. This is the heart of the mystery. The diaphanous waves exist wherever allowed, evolve and propagate in multiplex ways, and when we observe the electron, it is not a wave but a discrete entity, and it is not at a predetermined place but almost anywhere. We know only the chance or probability of where and when and how it will appear and do something observable.

We have a wraithlike quantum world of ghostly waves where all is fully determined and predictable. Yet, when we translate it into our observed world of sensible things and their events, we are limited to the concept of chance and the language of probability. What happens at the interface of the quantum world and the observed world may be this or may be that.

A single excited atom is a wavelike system that evolves continuously and predictably in a well-understood manner. To the observer it has a probability of decaying abruptly at any moment, but exactly when is never known. The deterministic precision of the quantum world contrasts with the fortuitous imprecision of the observed world. We are sometimes puzzled, though most of the time we are not bothered very much. Often, we deal with numerous atoms whose average behavior is predictable. Many atoms together behave in a continuous manner, and we can predict statistically how many will decay in each interval of time. Notice how the continuous and predictable behavior of many atoms in the observed world mimics the continuous and predictable behavior of single atoms in the wavelike quantum world. Why? Nobody knows.

Light and all radiation behave in this fashion as either waves or particles. The particles of radiation are known as photons. A wave of light strikes a target such as a photographic plate; the light is not absorbed smoothly and uniformly, as one might expect, but in discrete packages of energy—as photons—at haphazard points on the surface of the target. If the intensity of the light is made weaker and weaker, eventually a stage is reached at which we are able to detect and count the arriving individual photons. Again, this is very puzzling. The rippling wave is everywhere; we understand how it travels in space and how it is incident on the target, yet when we try to observe it, we detect discrete photons and know only the chance of where and when they will appear.

We must understand that atoms vibrate with rhythmic modes and are excited into states of various energies. As the modes of excitation evolve, waves of light are emitted and absorbed. The waves travel in space, and on arrival at the retina of the eye, other atoms absorb the incident waves. This vibrational give-and-take of atomic energy, in which atoms act individually and collectively, accounts for our visible world.

In the quantum world of potentiality, everything is becoming and never in an actual state of being. We calculate how the atomic waves and light waves evolve, and then we know in the observed world the probabilities with which atoms make transitions and photons are emitted and absorbed. The quantum world is deterministic and potential, the observed world statistical and actual.

Poets have yet to catch up with the antics of the atomic and subatomic world and rhapsodize on the most marvelous things of nature.

☐　☐　☐

Almost everybody has heard of the uncertainty principle and knows that it has something to do with the phantasmagoria of the atomic and subatomic world. Many feel tempted to shrug it off as physicists' sorcery. The uncertainty principle is easily stated: the more precisely we know where a moving particle is now, the less precisely we know where it was in the past and will be in the future. We are given a trade-off in precision that comes as a result of the wavelike properties of the quantum world.

The particle is distributed as a wave, and this wave tells us more or less what can be observed: how the particle moves (its velocity or, rather, its momentum) and where the particle is located in space. Both items of information cannot be known simultaneously with unlimited precision; there is a trade-off in the precision with which each may be known. The more we know of one, the less we know of the other, and this is a fundamental property of the physical universe.

Ordinary waves traveling on the surface of water or through the air have a similar, though simpler, form of uncertainty. Their frequency is unknowable with utmost precision. The longer the waves are observed, the more certain becomes their frequency. When the measurement lasts for a short time, the observed frequency is less certain; when the measurement lasts for a long time, the observed frequency is more certain.

Subatomic particles behave as waves, and their wavelike properties account for the remarkable uncertainty principle. A particle is like the ambisphaenic snake (of a forgotten author): you cannot know precisely from where it comes, and

> Before it starts you never know
> To what position it will go.

One way of presenting the uncertainty principle is in terms of energy. In the quantum world nature is like a generous bank that lends out energy free of interest. But all loans without exception must be repaid within a specified period of time. The larger the amount borrowed, the shorter the time it can be used before it must be returned. Everywhere energy is borrowed and repaid continually by all atomic and subatomic systems, and this adds to the intricacy of the dance.

The energy borrowed multiplied by the loan period cannot be less than a certain value determined by Max Planck's fundamental constant. By our standard the amounts borrowed are extremely small, and their smallness is undoubtedly a fortunate circumstance. Imagine what would happen if credit loans for the purchase of unlimited amounts of energy were free of interest for a human lifetime; New York City, or any city, would gladly go on a spree at the cost of beggaring future generations.

Enough energy can be borrowed to create a particle for a brief moment only. The energy loan is then withdrawn, the ledger balanced, and the particle vanishes. These short-lived ephemeral entities are called virtual particles; they live brief, ecstatic moments on borrowed energy and are the same as ordinary particles in all other respects. These will-o'-the-wisp creatures (*creature* means something created) are of considerable importance; they influence the natural states of atomic systems and among other things are responsible for gluing together the atomic nucleus and preventing it from flying apart.

There are a few incidental complications. Nature, for instance, does not lend electric charge, and a virtual electron must be escorted by its virtual antiparticle of opposite charge. The antiparticle of the electron is the positron, which is similar to the electron but has a positive electric charge. The charges of the two particles add to zero, and hence only energy is used to create them simultaneously as a pair. Constantly, energy is borrowed to create not only electrons and positrons, but also all other kinds of particles and their antiparticles. The whole of space is flooded with a sea of seething virtual particles, all popping in and out of existence in mind-reeling numbers. A million trillion trillion virtual electrons exist at any moment in the volume of a thimble.

This raises an obvious question: Why do we not see, hear, touch, taste, and smell them? The answer is that each must repay every bit of its borrowed energy and is not allowed to spend an iota of the loan to make itself known in our world of packaged and bartered energy. On its return to limbo it leaves behind not a vestige of its borrowed energy.

On rare occasions virtual particles succeed in finding from somewhere enough energy to pay off their debt; then, accompanied by antiparticles, they are released from the debtors' prison and are free to enter the real physical world. New particles of this kind are all around us, born of energetic cosmic rays that pour in from outer space or generated in high-energy accelerators used by physicists for the study of subatomic structure. This does not mean that the total number of particles is on the increase. When an electron and a positron cease to be virtual, the positron quickly annihilates with the electron or some nearby electron, releasing energy in the form of photons, and we are left with the same number of electrons and positrons as before.

To make all virtual particles real would require the utmost energy, vastly beyond what is available today in the physical universe. Yet long ago, in the early universe, there existed sufficient energy, and the multitudinous virtual particles were real and had their moment of glory.

The nucleus of the atom is itself a dance of waves to a rhythm twenty octaves higher than the electron waltz. This compact central region of the atom contains protons and neutrons, both much heavier than electrons. The proton has a positive charge, and the neutron has no electric charge. The nucleus also contains virtual particles, such as pions, which skip to and fro among the protons and neutrons and account for the strong forces that cement the nucleus together. The nucleus of the hydrogen atom is the simplest and lightest of all nuclei, consisting of a single proton only. The nuclei of other atoms have protons and neutrons in various combinations; for example, the helium nucleus has two protons and two neutrons, and the iron nucleus twenty-six protons and thirty neutrons.

In the atomic world energy exists in two forms—chemical and nuclear—and to understand this state of affairs we need not be atomic wizards.

Most chemical energy is released and absorbed when atoms are combined into molecules of various kinds. Thus chemical energy is absorbed when food is cooked, and the absorbed energy is used for rearranging the atoms in the carbohydrate and protein molecules. And whenever wood, coal, or oil is burned, some of the atoms are rearranged into new molecules such as carbon dioxide, and at the same time chemical energy is released and appears in the form of heat. Our biosphere on Earth is a system of orchestrated molecules continually interchanging energy. Some of this symphonic energy we redirect for our own use, which is a fine idea when not overdone. Like fishing the seas, if you take too much, the system breaks down, and you are left with nothing.

Sunlight is of paramount importance and bathes the Earth with photons of just the right range of energy for promoting the formation of organic molecular systems. The chemical energy stored in wood and in the fossil fuels of coal and oil came originally from sunlight, and sunlight derives from nuclear energy. Nuclear energy is released and absorbed whenever protons and neutrons are combined and rearranged into nuclei of different sorts.

The distinction between chemical and nuclear energy is this: the rearrangement of atoms in molecules involves the release and absorption of chemical energy, and the rearrangement of protons and neutrons in atomic nuclei involves the release and absorption of nuclear energy. An important difference between the two is that nuclear energy is doled out and gulped up in amounts about a million times greater.

Through my window I see the Sun. That shining orb, poised in the sky, is a titanic nuclear reactor. It is a star, a globe of hot gas held together by gravity, consisting mostly of hydrogen that is slowly being converted into helium. The Sun's surface has a temperature of 6000 degrees, and its center a temperature of about 10 million degrees. Because of the enormous interior temperature, hydrogen atoms in the Sun are stripped of their electrons, and the hot gas consists mostly of free protons and electrons moving around independently at high speed. In 1924 Arthur Eddington portrayed the scene in *The Internal Constitution of the Stars:*

> The inside of a star is a hurly-burly of atoms, electrons and aether waves. We have to call to aid the most recent discov-

eries of atomic physics to follow the intricacies of the dance. We started to explore the inside of a star; we soon find ourselves exploring the inside of an atom. Try to picture the tumult! Dishevelled atoms tear along at 50 miles a second with only a few tatters left of their elaborate cloaks of electrons torn from them in the scrimmage. The lost electrons are speeding a hundred times faster to find new resting places. Look out! there is nearly a collision as an electron approaches an atomic nucleus; but putting on speed it sweeps around it in a sharp curve. A thousand narrow shaves happen to the electron in 10^{-10} [one ten-billionth] of a second; sometimes there is a side-slip at the curve, but the electron still goes on with decreased or increased energy. Then comes a worse slip than usual; the electron is fairly caught and attached to the atom. Barely has the atom arranged the new scalp on its girdle when a quantum of aether waves runs into it. With a great explosion the electron is off again for further adventures. Elsewhere two of the atoms are meeting full tilt and rebounding, with further disaster to their scanty remains of vesture.

At the time Eddington wrote these words he did not know that the Sun consists mostly of hydrogen; his vivid picture, however, needs little alteration. In the same vein, he continued: "As we watch the scene we ask ourselves, Can this be the stately drama of stellar evolution? . . . The knockabout comedy of atomic physics is not very considerate towards our aesthetic ideals; but it is all a question of time-scale. The motions of the electrons are as harmonious as those of the stars but in a different scale of space and time, and the music of the spheres is being played on a keyboard 50 octaves higher."

Luminous stars like the Sun radiate energy into space for billions of years and have therefore a long-lasting internal source of energy. This source was unknown in Eddington's day, although nuclear energy of some kind was suspected.

Protons (the nuclei of hydrogen atoms) in the deep interior of the Sun move around in all directions at high speeds, continually encountering one another. Each collides about a trillion (a million million) times a second with other protons. But because protons are positively charged, they repel one another, and when any two rush to meet each other, they are pushed back and turned aside by their mutual repulsion. Protons in ordinary stars have, therefore, little chance of ever coming very close together. In the whole of the Sun not a single proton has enough energy to penetrate the electric repulsion barriers that keep protons apart.

Our picture of protons flying about like small, solid bodies is quite misleading. It overlooks the important fact that protons move about and interact with one another in a wavelike manner. Behaving as waves, like light feebly penetrating through a dark window, they occasionally filter through the repulsion barriers that normally separate them. About once every second each proton in the central region of the Sun succeeds in making a penetration and comes face to face with another proton. In these fleeting close encounters each brings into play its strong nuclear force. If that were the end of the story, it would also be the end of us. In one second only there would occur an immense release of energy, and the Sun would explode.

Life exists on Earth because protons are shy creatures. When brought face to face, they take a considerable time in deciding whether to like each other. Before their minds are made up they have moved apart and gone their separate ways. A similar thing happens to people in cities; they move about, encountering one another on the streets and in the subway, and sometimes one person meets another for a fleeting moment and feels a strong affinity. But in the hurry and bustle they turn aside and go their separate ways, perhaps never again to meet. There is a reservation between strangers that prevents them from making a snap decision and becoming intimate friends immediately. Protons have an equivalent inhibition, and their shyness and inability to make snap decisions is due to what is called the weak interaction.

The problem is this: no nucleus exists consisting of two protons only. Protons repel each other too strongly to form a nucleus. But a nucleus exists, consisting of one proton and one neutron; it is the deuteron, the atomic nucleus of heavy hydrogen. When two protons come close together and interact strongly, one of them has got to change into a neutron (by emitting a positron and a neutrino) so that both can be wedded into a deuteron. This switching of identity, of one proton changing into a neutron, while both are close together and strongly interacting, involves the week interaction that works extremely slowly. The probability of forming a deuteron during the brief encounter is extremely small. A proton in the central region of the Sun takes on the average 10 billion years to unite with another to form a deuteron. When this happens there is a tumultuous honeymoon and a release of nuclear energy.

Once a deuteron forms, it combines in 100,000 years with another deuteron to form a helium nucleus, and further energy is released. The nuclear energy unlocked by the conversion of hydrogen into helium maintains the Sun as a luminous body and supplies the energy radiated from its surface as sunlight. The conversion or transmutation of hydrogen into helium is a slow and continuous process, and after about 10 billion years the hydrogen in the center of the Sun is at last exhausted. The lifetime of hydrogen is roughly the luminous lifetime of the Sun as an ordinary star. When the Sun has consumed its hydrogen, which will occur in about 5 billion years time, it will swell into a red giant, then quickly turn into a white dwarf, "palely loitering" in the skies, with no further supply of nuclear energy.

Stars more massive than the Sun, after burning their hydrogen, become luminous stellar giants of even higher temperature and have access to further supplies of nuclear energy by burning helium into heavier elements, such as carbon and oxygen.

□   □   □

The biggest nuclear reactor in this part of the Galaxy is our genial lord and master the Sun, whose beneficent radiation pouring out into space derives from the release of nuclear energy. A technological dream of the modern age is to discover a way of burning hydrogen into helium, as in the Sun, and release energy on Earth in a controlled and steady fashion.

Ten billion years is much too long a time to wait, and fortunately the prolonged weak-interaction courtship between protons can be avoided. Heavy hydrogen whose atomic nuclei are already in the form of deuterons is moderately abundant, and enough exists in the oceans to meet the energy needs of humans for millions of years to come. All we need do is heat the heavy hydrogen to a temperature of about 100 million degrees; it will then burn to helium, and useful energy will be released without the long delay caused by the weak interaction. But so far we have not found how to do this in a way that liberates energy steadily and not explosively. We know, heaven help us, how to ignite heavy hydrogen explosively in the hydrogen bomb. But how to do it in a controlled manner for the benefit and not the ruin of mankind still eludes us, despite the sustained efforts of many scientists over the last three decades.

Already we have nuclear power reactors. From them we get electricity and the plutonium of nuclear weapons, and the nuclei of atoms have become the subject of sensational news. "I fear the Greeks even when they are bearing gifts," said Virgil. The gift of atoms by the Greeks has brought us hazardous radioactive wastes and turned our world into a nightmare of suicidal nuclear weapons.

The nuclear reactors we have at present do not obtain their energy from the fusion of hydrogen into helium, as in stars, but from the fission of uranium and plutonium into nuclei of lighter weight. Nuclear energy is obtained either by bringing together very light nuclei (fusion) or by breaking up very heavy nuclei (fission). The fission method is messy;

its ashes remain unavoidably radioactive for long periods of time, and we have as yet no foolproof way of disposing of them.

There is widespread concern, and many sections of the public believe that chemical energy in the form of coal is much more attractive than nuclear energy. I remember the great fogs of London, when I crept to school through the gloom of its choking atmosphere, guided by my hand trailing along railings and walls, scarcely able to see the ground underfoot. Every few months came the news of a fresh disaster at the coal mines. There was more radioactivity in the sooty air of London and other cities than is found nowadays in these places. Like everybody else, I am emotional on the subject of energy, and the prospect of returning to a coal-powered society when oil and natural gas run out is to me not in the least attractive.

We are beset by population problems that we in the West have helped to create by means of science, medicine, and technology, which science, medicine, and technology alone and unaided are now unable to solve.

On the one hand, atmospheric carbon dioxide is increasing—owing to the burning of fossil fuels and the destruction of primeval forests—which surely will not abate if we continue at the present rate. We are told that this will cause the climate to get warmer. On the other hand, atmospheric dust is increasing—owing to the farming of semi-arid lands to feed the human multitudes—which will also not abate with continual growth in population. Atmospheric dust shields the Earth's surface and tends to make the climate cooler. The two tug in opposite directions, with other known and unknown agents pulling in various and often uncertain directions, creating a state of biospheric tension. Possibly the climate will be jerked out of its present condition of global equilibrium. Pushed one way, and we might have deserts spreading everywhere; pushed the other way, and the mile-high glaciers might once again thunder down upon us. And acid rain as a bonus of chemical energy is poisoning the biosphere.

Other energy sources—tides, oceans, winds, rain, direct sunlight—cannot with our present technology supply the immense power needs of twentieth-century industrial society, nor can they for decades to come. That sublime product of nuclear energy, sunlight incident on the Earth's surface, is more than sufficient, if only we knew how to use sunbeams in heavy industry for feeding blast furnaces and steel mills and electric power stations and to meet the needs of transportation. The power we derive from sunlight and from geophysical sources at present is hardly sufficient to maintain a medieval standard of living for our vast twentieth-century human population.

Relentless circumstances have compelled us to make the Great Gamble. We are forced into a situation in which at all costs we must maintain the Western industrial society. To gain the time necessary to discover the know-how of using direct sunlight, and to learn how to break through to fusion, we must preserve Western society with its advanced industries and technologies. Whatever threatens the industrial society imperils the future of mankind. Oil and natural gas, our present mainstays, are fast running out, and in the decades ahead will quickly become inadequate. Coal will help us to keep going, as will fission reactors. Both are messy and part of the Great Gamble. Many persons will die, particularly from the effects of coal, but fewer than those killed in wars and on our roads.

If the Great Gamble fails, our way of life will inevitably falter as the power base melts away, and with the demise of our civilization will vanish the advanced technologies that some day might have harnessed direct sunlight and solved the fusion problem. On a sparsely populated globe (because billions will have died from hunger and disease) our descendents will probably live like our ancestors in slave-powered and serf-bound societies ruled by the despotic gods of a renascent mythic universe.

If the Great Gamble succeeds, despite the short-sighted efforts of many well-intentioned persons, we shall bring to Earth the power of stars and forever dispense with the messy and hazardous sources we have at present. At least our descendents will have the option of not reverting to ancestral ways of living.

☐ ☐ ☐

We try to explain our world of plants and flowerpots—of ordinary and sensible things—by delving into atoms and reaching out to the stars. Yet in this quest we find not the simplicity we seek, but utmost complexity that itself has urgent need of explanation.

The rich diversity of our environment breaks down into an assortment of millions of different kinds of molecules, which themselves break down into less than a hundred different atoms. At first glance, the atoms decompose into three different particles—electrons, neutrons, protons—and it must be admitted that thus far we have achieved considerable simplification. But now, as we delve deeper, seeking to understand more, there opens up a subterranean world of depthless mystery.

Electrons, protons, and neutrons are only three of the numerous kinds of particles now known. The electron is the most familiar example of the lightweight subatomic particles called leptons. At present there are six leptons: the electron and its neutrino, the muon and its neutrino, the tauon and its neutrino, and each has its antiparticle, making twelve leptons altogether. The proton and neutron are the familiar examples of the heavyweight subatomic particles called hadrons. The hadrons divide into two families—baryons and mesons—and hundreds of different specimens of both kinds have been discovered.

A few years ago the hadrons were thought to be among the ultimate constituents of the physical world. Now we attribute their properties to the existence of more fundamental subatomic particles of a yet deeper realm, which combine in various ways to form hadrons. These strange new creatures, called quarks by Murray Gell-Mann (from James Joyce's "Three quarks for Muster Mark!"), interact among one another with quixotic forces that fail to get weaker as their separating distances increase. When you try to tear quarks apart from one another, increasing their separating distances, their attractions remain strong, and the work performed in the attempt creates new quark combinations. Trying to separate two quarks is like trying to isolate the ends of a piece of string; if you pull hard enough, the string breaks, and you are left with two new ends. Trying to isolate the ends creates new ends, and trying to isolate the quarks creates new quarks. Quarks exist together in groups of two (forming mesons) and in groups of three (forming baryons) and cannot, so far as we know, be observed as isolated entities.

We have reached the point of postulating fundamental particles that, in principle, cannot be observed directly as isolated entities existing in their own right. They are beyond the reach of direct verification. This is something new in science, tantamount to postulating the mythical gods of long ago. Our incredulity is dismissed in much the same way as the incredulity of pagan atheists was dismissed on the grounds that gods are beyond direct verification and cannot be observed in person. The comparison is sufficiently apt for some theologians to sit up and take notice. If scientists can get away with it, why cannot they?

At the moment of writing there are six leptons; also six quarks (distinguished by the names up, down, charm, strange, top, and bottom), and each quark is dressed in one of three colors. When we add them all together with their antiparticles, and include the eight gluons that mediate between the quarks, we get fifty-six distinctly different subatomic particles. To this sum must be added the photon of the electromagnetic field, the elusive graviton of the gravitational field, the particles mediating in the weak interaction, and yet others too exotic to mention.

We began by seeking to understand things of sensible size, such as plants and flowerpots. We delved down, through the molecular realm with its DNA and other elaborate structures, to the atomic realm with its apparent order and simplicity. Beneath the atomic realm has opened up a subatomic world of amazing complexity that teems with strange particles of many kinds. For all we know this might not be the end of the search. Deeper realms may exist, consisting of even more exotic particles of even lusher variety. Though we tell ourselves that we are at last uncovering the ultimate secrets of nature, most of us, alas, have for-giveable moments of heretical misgiving. What good, pray, are the secrets of the subatomic world when they seem more mystifying than the comparatively sensible atoms they purport to explain? Can they account for plants and flowerpots any more than the remote and receding galaxies? Are we any nearer to answering John Hall, the seventeenth-century poet?

> If that this thing we call the world
> By chance on atoms was begot
> Which though in ceaseless motion whirled
> Yet weary not
> How doth it prove
> Thou art so fair and I in love?

We explain the world in terms of substructures and superstructures. Each structure has its specialized functions and is itself a thing that on closer examination consists of an activity of component structures. A seemingly endless vista reveals smaller and smaller systems. And by reaching out to the stars and galaxies we add components together, piling one on top of the other, creating a seemingly endless vista of larger and larger systems. We live in hope that limits are set to these vistas, that an ultimate goal exists at which a maximum of simplicity can be discovered. But we are not yet in sight of it. Is the goal to be found by proceeding in just the same old way by endless division and endless multiplication? It is conceivable and certainly not impossible that we have lost our way, that the goal is actually receding. Perhaps we have failed to recognize what is the goal; perhaps the ultimate limits of the small and the large converge and are indistinguishable from each other.

□   □   □

Diaphanous, otherworldly figures weave and twirl before our eyes to the melody of the atomic world, and as we watch, we ask, Can this be the dance that will rain death from the skies?

These are not the figures who will turn the keys and press the buttons, who will launch the missiles that will devastate the cities of Earth. No, these are not the figures in peril of being damned by all who have lived and might have lived.

There is another dance. The dance of human figures with tense faces, of men and women cavorting to a sociopathic throb of drums. I fear not the dance of atoms, only the dance of humans in a universe they themselves have made, which threatens to be the last universe in the history of mankind.

9

Fabric of Space and Time

We take space and time for granted. Normally they do not trouble us, yet whenever we think about them we become puzzled.

Space seems simple enough. Here it is, all around us, stretching away and spanning everything in the external world. We are surprised when told that people in other cultures have different ways of regarding space. What is there about it that can possibly be different? Edward Hall in *The Hidden Dimension* says, "there is no alternative to accepting the fact that people reared in different cultures live in different sensory worlds"—in other worlds of space. It seems that Arabs, Japanese, Hopi, and the people of other cultures have different styles of thought and modes of expression concerning arrangements and relations in space; they live in different mental worlds, in other worlds of space.

Time is much more puzzling. Here it is, stretching away and spanning everything in the past, present, and future. But unlike space it is not all around us. We experience time within ourselves, it seems, and cannot see it directly in the external world. Those intervals of minutes and hours on the face of a clock are actually intervals of space. A second cannot be displayed directly in pure form in the external world in the same way as a centimeter. This lack of objectivity about time greatly puzzled Robert Hooke in the seventeenth century: "I would query by what sense it is we come to be informed of time." Space is out there and obviously objective, yet time is in here and apparently subjective.

That other cultures have different ways of regarding time is not at all surprising. The Australian Aborigines believed they could withdraw into a dreamtime where the past with its ancestral figures coexisted with the present.

The ancients believed in the Wheel of Time, in the eternal return of the same patterns of events: the Sun rising and setting; the Moon waxing and waning; the seasons coming and going; the king dying and yet living; birth and death alternating in endless incarnations; nation triumphing over nation, catastrophe following catastrophe; wheels turning within wheels, cycles enfolding cycles; yuga replacing yuga, maha yuga replacing maha yuga, with the Days of Brahma numbered, though seemingly endless; and the gods, creating and destroying worlds, themselves doomed to die, tied to the vast and relentless Wheel of Time.

The cyclic view of time flourished in the Greco-Roman world and formed the basis of the Stoic philosophy and its message of fortitude in defiance of fate. The Mayas most of all were obsessed by the carousel of time; they believed their fantastic calendric computations ensured the periodic return of the time-carrying gods, and ritualistic and computational errors would terminate time by putting an end to their whirligig universe.

"What, then, is time?" asked Saint Augustine in the *Confessions*. "If no one asks me, I know what it is. If I wish to explain what it is to him who asks me, I do not know." Time, he thought, extends from the past we remember, through the present we now know, to the future we foresee. Augustine's view of time was much the same as the commonsense view of today. But it caused him much distress. His perplexity, expressed by Austin Dobson in *The Paradox of Time*, is with us still:

> *Time goes, you say? Ah no!*
> *Alas, Time stays, we go.*

"Time, like an ever-rolling stream, bears all its sons away," says the hymn. We think of time as a river, carrying us forward or flowing by us, moving from the past to the future. In the *Principia* Isaac Newton wrote, "time, of itself, and from its own nature, flows equably without relation to anything external." Time flows, said Newton, and we tend to agree. Hooke, ever at odds with Newton, was not so sure, and wanted to know where time is and how we apprehend it.

Augustine searched his soul, Newton got down to business, yet both said much the same: not the Wheel of Time but the River of Time. Both regarded time as similar to space, as a one-dimensional extension in the external world, through which we move from the past toward the future. The "now" with its memory of the past, its vivid awareness of the present, and its anticipation of the future moves through time like a bead sliding on a wire. "When we evoke time," said Henri Bergson, "it is space that answers the call." This view of spacelike time, now common in Western society, entails the notion of movement through time.

We move through time or time flows past us. In either case, we move relative to time. The notion of movement in time, when we think about it, is quite absurd. It compels us to ask: At what speed do we move in time? Already we have used time once, and now it must be used again to tell us the rate at which we move in time. Though the notion is thoroughly absurd, we use it continually, because we have nothing better.

If the "now" moves through time, there must be a second time, argued John Dunne in *An Experiment with Time*. But in this second time we have motion also, and hence there exists a third time, implying a fourth, and then a fifth, and so on without end. Most persons find Dunne's notion of serial time unappealing. His argument, however, has the virtue of making clear the consequence of believing that we have motion in time.

In what follows I take the common view that time is spacelike and extends one-dimensionally from the past to the future. At first I shall not be overly concerned that this representation poses an unsolved riddle.

□   □   □

Space in the Newtonian universe existed in its own right. Distances between bodies were relative, but space itself was absolute. Unlike the Newtonians, the Cartesians followed Aristotle and thought that space could not exist by itself as a vacuum. To them, space was nothing more than a property of matter, and where there was no matter, there could be no space. The controversy over Aristotelian clothed space and Newtonian unclothed space persisted into the eighteenth century, and Newtonian space finally triumphed. The notion of time flowing "equably without relation to anything external" caused surprisingly little controversy. We inherit that lack of critical concern, and while rejecting the nonprogressive futility of the Wheel of Time, we are blind to the progressive fatuity of the River of Time.

The Newtonian universe dominated the eighteenth and nineteenth centuries. Inhabitants of this universe shared the same public space and the same public time. All persons agreed on their measured intervals of space between places and their measured intervals of time between happenings. This is the commonsense world we live in that accommodates the furniture of the biological and social sciences. Twentieth-century physics wrenches the mind by rejecting it.

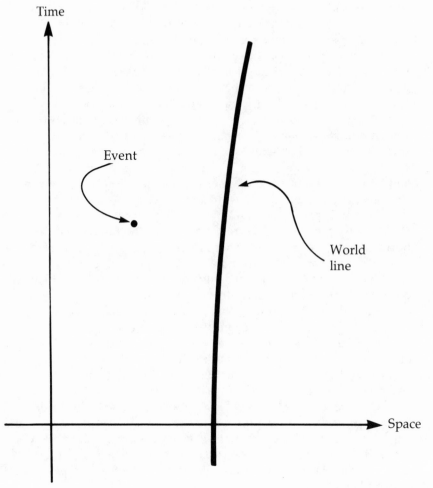

A space-and-time diagram, showing an event and a world line.

A meter stick that is held up, visible to all, is a measured interval of space. I cannot hold up an interval of time (even if I knew how, I doubt if we could see it), yet I can easily demonstrate intervals of time by asking you to listen to the ticks of a clock or count the swings of a pendulum. Time is continuous; its intervals are measurable, and hence time is spacelike. We talk of so many centimeters, meters, or kilometers from one place to another, of so many seconds, minutes, or hours from one occurrence to another, and our lives are regulated within the quantifiable domains of public space and public time.

Intervals of time may be combined with intervals of space. We combine them repeatedly when speaking of the speeds of bodies; 8 kilometers an hour, a walking speed, is about 2.5 yards a second; 55 miles an hour, almost 25 meters a second, is now our maximum driving speed.

Scholars in the late Middle Ages at the universities of Oxford and Paris defined speed and acceleration. Speed is the distance traveled in space in an interval of time. Acceleration is increase in speed, also in an interval of time. These seminal ideas of quantified motion, made clear and vivid with graphs and diagrams by medieval scholars, formed the first stepping stone to the Newtonian laws of motion.

We can display space and time in a diagram. Here is a vertical line representing time stretching from the past to the future, and here is a horizontal line representing space stretching from place to place. Of course, we cannot display the three dimensions of space on a blackboard or a sheet of paper, and for our purpose it is sufficient to display only one dimension. All this was understood five centuries ago, and we have encountered so far nothing in this discussion to be alarmed about.

In the space-and-time diagram a point represents an event. An event is something at a place in space at an instant in time, such as the blink of an eye or the flash of a firefly. Events generally are the things we observe and are represented by points or small regions in the space-and-time diagram.

At a public lecture on *Space and Time* at Cologne in 1908 Hermann Minkowski said, "Nobody has ever noticed a place except at a time, and a time except at a place." We notice events at specific places at specific times. But have you ever wondered why an object may be observed at the same place in space at two instants of time but not at the same instant at two places?

The birth of a child is an event. The child grows, experiences many events, then dies, and death is the last event. These events from birth to death when strung together form a line in the space-and-time diagram. This life line, called a world line, shows the position in space of the person at each moment of time. All things that endure, such as atoms, bacteria, human beings, and stars, are represented by world lines in the space-and-time diagram. Objects at rest relative to one another have parallel world lines; objects in relative motion have world lines inclined to one another. Again, there is nothing to be alarmed about.

What the Newtonians said, and everyone agreed, was that between any two events the measured intervals of space and time are the same for everybody. If one person said that the separation between two events was so many feet in space and so many seconds in time, all other persons making the same measurements obtained the same results. They were in total agreement on their measurements. Even hypothetical creatures moving at very high speed in spaceships obtained the same results. This seemingly logical outlook changed in the first decade of the twentieth century because of a revolutionary theory, later known as the special theory of relativity.

□ □ □

The electromagnetic theory developed by James Clerk Maxwell in the 1860s was very puzzling. This elegant and powerful theory, which unified electricity and magnetism, showed that waves of light travel at a definite speed through empty space. Light has a speed of 300,000 kilometers a second, and we now know that all electromagnetic radiation—radio waves to X-rays—travels at this speed.

But empty space is a sort of nothing and just a vacuum. How, then, can waves of light move at a definite and constant speed relative to nothing? It looked in the nineteenth century as if perhaps Descartes were right after all, and space could not exist unless clothed with a material medium relative to which light had definite speed. Many efforts were made to conjure up a light-transmitting medium consisting of an undulating ether. It became the fashion to speak of light and other forms of electromagnetic radiation as ether waves.

If light moves at constant speed in the ether and if the Earth moves through the ether while revolving around the Sun, then by measuring the speed of light, it should be possible to detect the Earth's motion through the ether. Using utmost precision, Albert Michelson and Edward Morley in 1887 attempted to detect the motion of the Earth by measuring the speed of light. To their surprise they found that the Earth's motion defied detection, and the speed of light measured on Earth is the same in all directions at all times of the year.

Consider the light received from a distant star. Let us suppose at first the Earth moves away from the star; 6 months later, after swinging around the Sun, the Earth moves in the opposite direction toward the star. Yet on both occasions, when the Earth has motion away from and toward the star, the observed speed of the starlight is the same. The speed, moreover, remains unchanged throughout the year. The implication, later confirmed, is that the speed of light from all sources is constant for all observers, no matter how fast the sources and observers move relative to one another.

In the declining years of the nineteenth century George FitzGerald of Ireland and Hendrik Lorentz of Holland tried to get around the problem by supposing that matter and clocks in motion altered in such a way as to maintain the observed constancy of the speed of light. Everything moving through the ether had its size adjusted so that light appeared to have constant speed. According to this theory, the laws of nature conspired for unknown reasons to create the impression that light, moving through the ether, had constant speed for all observers. This makeshift theory, though not very elegant, was helpful and suggestive to Albert Einstein.

Maxwell's electromagnetic theory and the puzzling constancy of the speed of light delivered the deathblow to the Newtonian universe. In 1905 Einstein threw away the ether, then advanced the theory of relativity that has revolutionized our understanding of space and time. Instead of trying to explain Einstein's algebraic treatment, I shall use a simpler approach that offers greater conceptual insight.

□　　□　　□

Charles Hinton, a gifted educator and scientist, published in 1887 the thought-provoking book *What Is the Fourth Dimension?* He proposed that the world is actually four-dimensional, not three-dimensional, and the fourth dimension is time. He argued, unlike others, that our world is literally, not figuratively, four-dimensional.

The world lines in the space-and-time diagram, explained Hinton, are not just a convenient way of illustrating motion, but are representations of actual objects in a physical world of unified space and time. As we move along our world lines we see revealed a changing three-dimensional world of space. We travel along our world lines in a four-dimensional world like trains, and we see an ever-changing countryside of three-dimensional space.

H. G. Wells, inspired by Hinton, wrote *The Time Machine* 8 years later. He tried to explain to a different audience the idea of time as a fourth dimension. "For instance," wrote Wells, "here is a portrait of a man at eight years old, another at fifteen, another at seventeen, another at twenty-three, and so on. All these are evidently sections, as it were, Three-dimensional representations of his Four-dimensional being, which

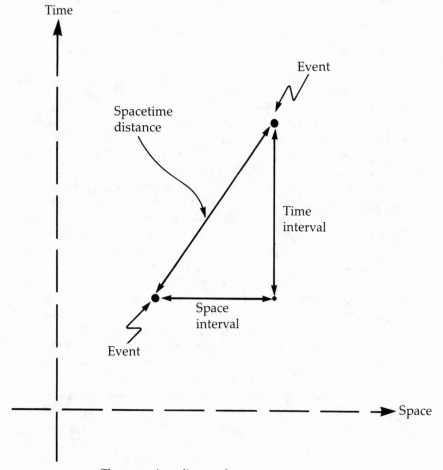

The spacetime distance between two events.

is a fixed and unalterable thing." By portrait he meant the actual person, not just a two-dimensional picture.

Einstein advanced the theory of relativity, and Minkowski, his former teacher, explained in 1908 what the theory meant in terms of the space-and-time diagram. Minkowski said: "The views of space and time which I shall lay before you have sprung from the soil of experimental physics, and therein lies their strength. They are radical. Henceforth space by itself, and time by itself, are doomed to fade away into mere shadows, and only a kind of union of the two will preserve an independent reality."

Up till then scientists had regarded the idea of a four-dimensional world as little more than a convenient graphical way of representing motion in space and time. Minkowski showed that space and time as a result of Einstein's theory are actually fused together into a world of spacetime. The intrinsic structure of spacetime accounts for the constancy of the speed of light for all observers.

The theory of relativity replaced the public space and public time of the Newtonian universe with a unified public spacetime. Intervals of space and intervals of time between events are no longer common property; instead, intervals of spacetime and the speed of light are the things on which we all agree. The length of the spacetime interval between any two events is the same for everybody. (Notice how the invariant Newtonian intervals of space and time are replaced with two new invariants: the speed of light and the spacetime interval.)

In the physical universe of today we all agree—no matter how fast we move relative to one another—on the measured value of the spacetime distance between any two events and on the measured value of the speed of light. We do not agree on our measurements in space and time, only on a combination of these measurements that gives the spacetime distance. This amazing change in outlook has been forced on us by the discovery that space and time are not independent public domains. You and I share the same spacetime, but my space and your space and my time and your time are the same only when we are at rest relative to each other. Spacetime is the new public domain, and within it we have our own worlds of space and time. Einstein and Leopold Infeld in *The Evolution of Physics* wrote: "The relativity theory arose from necessity, from serious and deep contradictions in the old theory from which there seemed no escape. The strength of the new theory lies in the consistency and simplicity with which it solves all these difficulties, using only a few assumptions. . . . The old mechanics is valid for small velocities and forms the limiting case of the new one."

The relativity theory achieves consistency and simplicity at the cost of a fundamental change in outlook that is rarely explained consistently and simply to the world at large.

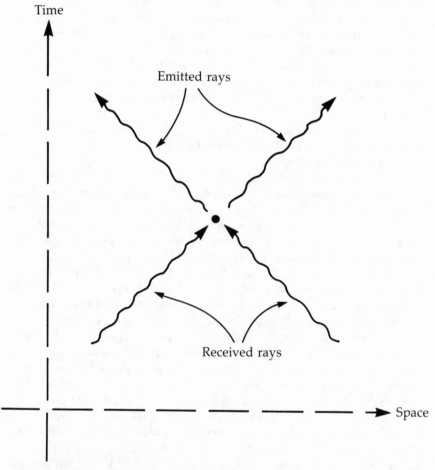

The emitted and received light rays at an event. Each light ray travels in 1 second a distance of 1 light-second.

Light in 1 second travels 300,000 kilometers, roughly the distance to the Moon, and this distance is called a light-second. The distance to the Sun is 500 light-seconds. Light from the Sun takes 500 seconds to reach us, and we see the Sun as it was 500 seconds ago. Light-travel time is a convenient way of measuring large distances, and it has the advantage that we know how far we look back into the past when observing a distant body. Nearby stars are at distances of roughly 10 light-years, or 100 trillion kilometers, and we see them as they were 10 years ago; nearby galaxies are at distances of roughly 10 million light-years, or 100 billion billion kilometers, and we see them as they were 10 million years ago.

Let us measure intervals of space in light-seconds and intervals of time in seconds. (We could also use light-years as intervals of space and years as intervals of time.)

Our space-and-time diagram has become a spacetime diagram. At each event we may show light rays coming in from the past and light rays going out into the future. In 1 second these rays travel a distance of 1 light-second; they are, therefore, inclined at angles of 45 degrees, as shown in the figure. The rays are always inclined at 45 degrees, because the speed of light is the same everywhere for everybody.

What is the theory of relativity? It is simply this: Each world line decomposes spacetime into its own space and its own time. The time that belongs to a world line is measured along that world line. Time can be measured in any way we wish, by a watch, by an atomic clock, or by just counting heartbeats. Your time is measured along your world line by your clock, and my time is measured along my world line by my clock. The time that elapses between birth and death is the length of a person's world line. When things have motion relative to one another and their world lines are not parallel, they do not share the same time.

Each world line has its own space as well as its own time. The space that belongs to a world line is always perpendicular to that world line. When things have relative motion and their world lines are not parallel, they do not share the same space. On a sheet of paper we can only show space perpendicular to the vertical world lines; inclined world lines also have their spaces perpendicular, but unfortunately this cannot be shown in the same diagram, because of the strange geometry of spacetime that I shall come to shortly.

If you rush past me, dashing in through one door and out through the other, your space and your time are not quite the same as mine. Remember, time has been spatialized. Your time contains part of my time and some of my space, and your space contains part of my space

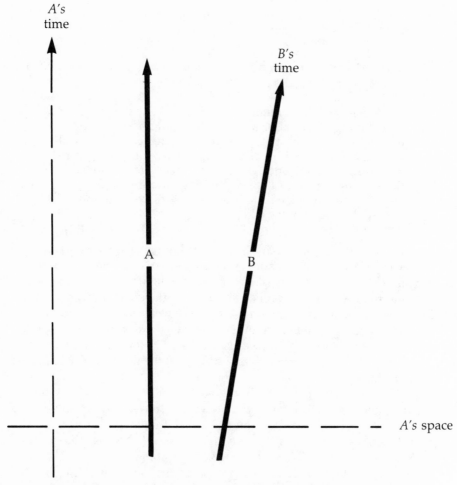

A's
time

B's
time

A

B

A's space

In the spacetime of special relativity time is measured along a world line. Space is measured perpendicular to a world line, though this cannot be shown for all world lines in such a diagram as this.

and some of my time. And vice versa, my time contains part of your time and some of your space, and my space contains part of your space and some of your time. We share the same spacetime, but not the same space and time, and our world lines determine our different worlds of space and time.

□   □   □

In *The Mathematical Theory of Relativity* Einstein drew attention in 1911 to a remarkable aspect of special relativity:

> If we place a living organism in a box . . . we could arrange that the organism, after an arbitrarily lengthy flight, be returned to its original spot in a scarcely altered condition, while corresponding organisms which had remained in their original positions had long since given way to new generations. For the moving organism the lengthy time of the journey was a mere instant, provided the motion took place with approximately the speed of light.

This sums up the clock paradox, otherwise known as the twin paradox. One twin stays at home on Earth, and the other goes off on a long journey in a spaceship that travels close to the speed of light. Let us call the stay-at-home twin *A*, short for Albert, and the gadabout twin *B*, short for Bertha. After many years *B* returns from her travels, and it is immediately apparent that she is much younger than her twin brother *A* who has stayed at home. Age in both cases is measured in the same way: by the clocks they carry, the number of gray hairs, and the number of heartbeats since birth.

Albert feels cheated and is indignant. "It's not fair! We were born together, and now look at you—you are years younger." To which Bertha replies, "But we are not the same age. You are older, you have slept more times, ate more meals, and read more books." Their age differences is real, and the result of the geometry of spacetime.

It is important to realize that in spacetime a straight world line is not the shortest distance between two events. Many of the surprising results of relativity spring from this fact alone.

Most of us are familiar with the Pythagorean theorem: The square of the hypotenuse of a right triangle is equal to the sum of the squares of its two sides. Here is a triangle in spacetime. Common sense insists that the hypotenuse of this triangle must be longer than either of the two sides. But common sense deceives us, because the Pythagorean theorem does not apply to spacetime. The geometry of spacetime is not

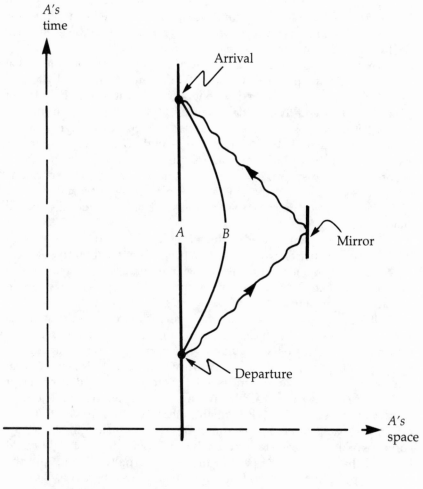

The stay-at-home twin is *A*, and the gad-about twin is *B*. In spacetime *B* travels a shorter distance than *A*.

the same as that of ordinary space. Triangles drawn on a blackboard or a sheet of paper fail to illustrate the properties of spacetime.

The distance between two events in spacetime is equal to the time interval squared *minus* the space interval squared. This minus rather than plus was Minkowski's great discovery. The Pythagorean theorem is different in spacetime. When we observe two events: always your time interval squared minus your space interval squared equals my time interval squared minus my space interval squared. This is true even when we live in different worlds of space and time.

We can now answer a simple and very important question: What is the distance in spacetime between two events connected by a light ray? A ray of light is emitted by one atom and absorbed by another atom. We are asked to find the distance in spacetime between the two atoms. Let us suppose that the ray travels for 1 second between emission by the first atom and absorption by the second. The distance in time is 1 second, and the distance in space is, therefore, 1 light-second. If we square each, then subtract, we get the result: $1 - 1 = 0$. The spacetime distance is thus zero. If the atoms are separated by x seconds in time and x light-seconds in space, we get the same answer: zero distance in spacetime.

Spacetime is constructed in such a way that the distance traveled by light rays is always zero. Light rays from distant stars hurry at great speed for long periods of time across wide gulfs of space and yet travel no distance whatever in spacetime. In the world of spacetime we are in contact with the stars.

Albert stays at home and his world line is more-or-less straight, as shown in the figure. Bertha leaves on her long journey, then eventually returns, and her world line is curved—it goes out and comes back again. Suppose at the moment of her departure a pulse of light is emitted from the Earth. Suppose that this pulse of light is reflected by a distant mirror and is received back on Earth at the moment of her arrival. The total distance traveled by the pulse of light, measured from the Earth to the mirror and back, is zero in spacetime. Obviously, the length of Bertha's curved world is somewhere between the length of Albert's straight world line and the length of the path of the reflected pulse of light. But the reflected pulse travels zero distance in spacetime. Therefore, Bertha's curved world line is shorter than Albert's straight world line. In spacetime curved world lines are always shorter than straight world lines: shorter, even though on a blackboard or a sheet of paper they look longer.

Time is measured along a world line. The amount of time that elapses between two events on a world line is the length of the world line between these events. Bertha goes off on her travels, and her curved world line is shorter than Albert's straight world line. The faster she travels, the more her world line approaches that of the light ray, and the shorter the time she spends on the journey. Her age when she departs is the same as Albert's and is less than Albert's when she returns. If she could travel as fast as a ray of light, her journey would take no time as measured by her clock. In one heartbeat she could traverse the universe.

The strange and surprising properties of spacetime are the result of the way space and time are fused together. When we combine the squared intervals of time and space we must subtract one from the other. As a result, light rays have zero length in spacetime, and curved world lines are shorter than straight world lines. With these simple facts in mind, the algebra of special relativity formulated by Einstein becomes smooth sailing. You, too, can become a relativist with only high-school algebra.

The effects of relativity are not apparent in ordinary life, because none of us travels very fast compared with the speed of light. In the laboratory, however, the effects are observed constantly. For example, an unstable particle known as the muon decays into an electron and a neutrino in a time of about one-millionth of a second. The muon's short lifetime is measured along its world line. Suppose the muon travels close to the speed of light. In the Newtonian universe we would expect it to decay after traveling about 300 meters, a distance equal to one-millionth of a light-second. But in our physical universe it travels much further. At a speed 99.5 percent that of light the muon goes about 3000 meters before decaying.

Imagine a spaceship capable of accelerating and decelerating at one g (equal to the acceleration at Earth's surface). Travelers inside this spaceship experience a force equal to their weight on Earth, owing to the constant change of velocity. The travelers can arrive at and land on a planet in the Andromeda galaxy in only 30 years of their time. Andromeda is the large, nearby galaxy at a distance of 2 million light-years. For the travelers the total journey there and back lasts 60 years in their time. If they start when young they can return in old age. On their return, however, they will find that the Earth has aged by 4 million years.

☐　☐　☐

Why does time flow from the past to the future? What determines the direction of the River of Time? This is the arrow-of-time riddle.

Our space-and-time diagrams fail to tell us which is the future and which the past. We could turn the diagrams upside down, and they would still be much the same. A space-and-time diagram, with the top labeled future and the bottom labeled past, contains little or nothing to prevent us from turning the diagram upside down so that the future becomes the past and the past the future. Yet in ordinary life the past and future are very different, and there is no possibility whatever of confusing the two. Daylight does not begin at sunset and end at sunrise, rivers do not flow from seas to mountains, and life does not begin in the grave and end in the cradle.

Most laws of physics cannot distinguish the past from the future and lack what Arthur Eddington called the arrow of time. The arrow-of-time riddle can be tackled to some extent in much the same way as the problem of deciding which way up to hang a picture on the wall. An examination of the brush marks and small dabs of paint fails to reveal which is the top and which the bottom of the picture. We must stand back and look at the shapes of larger regions. Much the same applies to the physical world.

On the microscopic scale there is very little in the physical world that determines the arrow of time. Two particles rush together, collide, and then rush away; when the arrow of time is reversed, they again rush together, collide, and rush away. What happens one way can happen the other. But on a much larger scale—that of plants and flower-pots—things are very different. A hot cup of coffee grows cold; an ice cube from the refrigerator melts; a drop of ink in a glass of water diffuses and disappears. Plants grow from seeds; heat travels from hot regions to cooler regions; energy cascades into more-dispersed and less-accessible states; organization gives way to disorganization; and the direction of time is unmistakable.

This familiar behavior of the ordinary world, which fixes the arrow of time, is governed to a large extent by the laws of thermodynamics. These laws, often adumbrated in a way calculated to confuse the non-scientific half of C. P. Snow's "two cultures," have been summarized by an unknown author as follows: "You cannot win" (cannot get something for nothing, because energy is conserved), "you cannot break even" (cannot keep on using the same bit of energy, because it cascades into less-useful states, and entropy increases), and "you cannot get out of the game" (cannot escape, because temperature has an absolute zero that is unattainable). With these laws and their large-scale effects we can usually determine which is the past and future in the space-and-time diagram.

Even so, the arrow of time is still not fully understood, and many perplexing aspects remain unresolved. It has been suggested that the direction of time for each world line is perhaps ultimately determined by the physical universe as a whole. According to this idea, the shapes and figures are not sufficient to show how to hang the picture on the wall—we must also be guided by the picture frame.

All around us is a world in action. From moment to moment and year to year things constantly change. We occupy a tumultuous world that is in a perpetual state of Heraclitean flux. Yet in the world of space-time there is no action, nothing changes, and all is at peace and rest in a Parmenidean stillness. In the tumult of our private worlds of space and time everything changes; in the stillness of the public world of spacetime nothing changes.

Things are displayed in spacetime in the form of events, world lines, and light rays. All that exists is there, unhidden and on show, depicted in an unchanging and tenseless manner. All that has been and all that is and and all that will be *is*.

Spacetime is like a crystal ball. Every little detail of the universe throughout spacetime is on display to the percipient fortune-teller: "My dear, you have had an unhappy childhood, you are worried about your job, but do not worry, you will soon receive a letter, and will then go on a long journey, meet a tall, dark man, have two children, and live happily in another country, and die in old age."

Each thread in the fabric of spacetime follows an appointed path. No surprises, no secrets, for all is fully disclosed and on display. Death does not hide from sight but is depicted as plain as birth. Maurice Hare in *Limerick* wrote,

> There once was a man who said "Damn!
> It is borne in upon me I am
> An engine that moves
> In predestinate grooves,
> I'm not even a bus, I'm a tram."

Spacetime is the world of fate in which every detail of life is on display and unalterably fixed. Fate is the enemy of free will.

Believers in fate hold that all is unalterably ordained. Whatever I do, whether I plan or do not plan, strive or do not strive, serves no purpose and accomplishes nothing more that what already is timelessly ordained. I can alter nothing, and as Milton said, "what I will is Fate." Human beings have limited vision, thus accounting for their illusion of free will; the gods alone can peer into the crystal ball of unalterable spacetime.

Meddling with fate and altering the cosmic design offend the gods of the mythic universe. The original sin of eating forbidden fruit was an act in defiance of fate. It was the first act of human free will and the worst sin, for it transgressed against God's will.

□ □ □

Determinism also is the enemy of free will. The old debate of determinism versus freedom of will continues to this day with undiminished vigor.

The doctrine of determinism declares that everything has its cause and nothing is arbitrary. What has happened determines what happens and will happen. Whatever happens has its reason and explanation; we seek the necessary reason and discover the sufficient explanation. Everything ultimately is rational, and nothing, in principle, inexplicable. Even human behavior is determined. A world of predetermined events allows no room for pure chance and free choice.

Imagine a world in which nothing had predestinate grooves and nothing could be predicted. It would be irrational. No person could live in it, and no society has ever devised such a universe.

Things are determined because we believe they are the consequence of causes and the cause of consequences. We may not always know the causes and consequences, but we believe nonetheless that they exist, have always existed, and will always exist. We cannot predict the weather next year on a certain day, but we are confident that whatever happens will have its cause. Physicists talk of the meaning of the uncertainty principle, but nobody takes them seriously; plain for all to see is the give and take of cause and effect, without which intelligent life and even physics would be impossible.

Determinism springs from the deep-rooted belief that the universe is rational and governed by intelligible laws. Spacetime offers a sort of godlike percipience of a fully displayed universe. We must be cautious, however, and not jump to the conclusion that this means the universe is therefore deterministic. The universe could be irrational, with every happening random, uncoordinated, and ungoverned by laws, and yet still be displayed as it was, is, and will be—as it *is*. To make the spacetime picture deterministic we must show how its various parts interrelate and its events coordinate.

Freedom of will is the conviction that we as individuals have some control over our lives and are not the sport of merciless fate and the helpless victims of inflexible laws. All scientific, philosophical, and theological theories that explain how things work are in conflict with our personal awareness of free will. As Dr. Johnson said, "All theory is against the freedom of the will; all experience for it."

Belief in free will flies in the face of contemporary wisdom and is the essence of the Pelagian heresy (discussed in Chapter 14, *The Design of the Universe*). Saint Augustine in the late fourth and early fifth centuries was the architect of a deterministic theism. Nothing acted freely, and everything conformed to a grand design. Freedom of will contradicted the omnipotent will of God, and human beings followed their predestinate grooves as ordained by God. "Give me what thou commandest, and command what thou wilt," said Augustine in his *Confessions*. The cost of a rational, explicable universe—any rational, explicable universe—is the loss of free will. Freedom of will becomes an illusion of our deceived wits and senses.

But Pelagius, a British monk of the same period, preferred a less-rational and more-mystical universe, in which not the will of God but the will of human beings accounted for human sin. Pelagius rejected original sin and was condemned as a heretic. All who claim freedom of will and deny the determinism of the universe in which they live are guilty of the Pelagian heresy. I am myself a Pelagian heretic.

□　　□　　□

Conceivably there is something deficient in our ideas of time. The Wheel of Time stresses the act of becoming, the River of Time stresses the state of being, and neither accounts adequately for the time that we experience. Time wheeling is not enough, time flowing is not enough, but both have complementary aspects.

The *now* is always with us. It embodies our conscious awareness of time, and the time we experience consists of states of being and acts of becoming.

In one sense we are aware of time as a state of being throughout which things are diversified. From this point of view the now embraces all time—the past, present, and future—in a state of being. This is the aspect of time that has been spatialized and woven into the fabric of spacetime. But in another sense we are also aware of time as an act of becoming, of one state of being flowing and wheeling into another state of being, of one vista of past, present, and future dissolving and re-forming into another vista of past, present, and future. The now of today with its past, present, and future is different from the now of yesterday with its past, present, and future. The tapestry of being in each act of becoming is rewoven. This aspect of time defies spatial representation. It has been omitted from the physical universe because we have so far not learned how to express it either linguistically or mathematically. To condemn the act of becoming as an illusion oversimplifies the world in which we live.

In the Wheel of Time everything transforms in acts of becoming, and nothing endures in states of being. With time forever wheeling, acts of becoming are vivid and real; states of being are evanescent and unreal.

Whenever we try to isolate a state of being, it slips away, eludes us, and at best is grasped as a monotony of dissolving forms.

In the River of Time we have a clear state of being, consisting of the past, present, and future. This is the aspect of time that we spatialize. But the act of becoming defies spatial description and creates the motion-in-time and the arrow-of-time riddles. The now and its awareness of the past, present, and future has become a segment of time that paradoxically moves unidirectionally in time. If we think of ourselves as world lines in spacetime, the now vanishes, and when we try to put it back with its awareness of becoming, we think of it as a segment of time moving along a world line like a spark along a fuse. But in a state of being, with its spatialized time, nothing changes, and the now cannot move along a world line. Spatialized time fails to represent our awareness of the transience of things becoming. Time that supposedly flows cannot flow, and when we try to grasp the act of becoming, it melts away, and we are left with only a paradoxical movement in time.

In the last chapter of this book (Learned Ignorance) I shall comment further on the nature of time. To tell the truth, I have nothing much to add, only to point out that a thorough-going theory of time that combines the aspects of being and becoming may have unexpected consequences.

If we cannot put the now with its act of becoming into the physical universe, then it seems safe to say that we cannot put consciousness and its awareness of free will into it either. We have failed to represent in the physical universe even the rudest aspects of ourselves as experiencing individuals. Possibly the next major step in the design of universe will be the discovery of a more sophisticated way of representing time.

10

Nearer to the Heart's Desire

We have a picture of seamless spacetime projecting into the space and time of each observer's world line. Though elegant and economic, it differs little from the Newtonian picture in one sense. Space in the Newtonian scheme was just a sort of nothing spanning everything, and time also a sort of nothing in which objects endured. Both came together in the theory of special relativity to form an expanse of spacetime that, again, was just a sort of nothing.

Then in 1916 Einstein advanced the theory of general relativity, and the picture changed once more. Spacetime lost its state of nothingness and acquired a tangible physical reality. Gravity ceased to be an astral force acting mysteriously at a distance and became a property of curved, dynamic spacetime.

Spacetime itself in the new scheme guides the heavenly bodies, and the old astrologically inspired action at a distance turns out to be the curvature of spacetime. We now have a palpable spacetime that pulls and pushes, that transmits shivers and shakes at the speed of light. We cannot eat spacetime any more than air, but it can be hit, and will hit back, and can eat us if we stray too close to a black hole. Spacetime in general relativity springs to life and becomes an active participant in the physical universe.

□ □ □

Something curious and rather interesting in the Newtonian universe points in the direction of general relativity. It led Einstein to his monumental theory.

On Earth we feel the pull of gravity. In an accelerating vehicle we feel a force that is very much like the pull of gravity. If we wish to contemplate nature undisturbed by the pull of gravity, we may follow Arthur Eddington's advice, "take a leap over a precipice," and enter into a state of free fall. Here in a nutshell are the essential ingredients of the equivalence principle that leads to general relativity. Let me try to explain.

Imagine a spaceship in the depths of space equipped with all kinds of apparatus. The spaceship has no windows, and the experimenters inside cannot see what happens in the outside world. We suppose the spaceship at first is far from any star and undisturbed by gravity. It moves freely at constant velocity. With their various instruments, the experimenters find themselves unable to determine how fast and in what direction their windowless spaceship moves. To them it seems motionless. A ball, for instance, floats above the floor and remains stationary inside the moving spaceship.

Any experiment performed inside the spaceship yields results always in conformity with special relativity. Gravity is absent, and spacetime is uncurved and unremarkable.

The spaceship eventually approaches a star, swings around the star in a curved orbit, and moves away in a new direction. While this happens the ball continues to float above the floor and remains stationary inside the spaceship. (Complications owing to tidal forces, or gravity variations within the spaceship, need not detain us in this discussion.) Tirelessly, the experimenters perform experiments with their various instruments and remain totally unaware of the gravitational pull of the nearby passing star.

The spaceship moves freely. It is a free-falling system following a trajectory of such a kind that the force due to its acceleration always cancels exactly the gravitational force due to the star. This simple and quite remarkable state of affairs lies at the heart of the Newtonian scheme of motion and gravity.

Perhaps you have difficulty believing that the force of acceleration cancels gravity exactly in a state of free fall. I shall not ask you to jump over a precipice to convince yourself of its truth; you might only find that air is a very resistive medium. Astronauts while orbiting about the Earth in free fall have shown us on television how objects float in a weightless state inside their space vehicles. The astronauts themselves experience this weightless state, because gravity is nonexistent in all free-falling systems. Spacecraft with idle rockets move in free-fall orbits that annihilate local gravity.

Our diligent experimenters in the depths of space remain under the impression that their windowless spaceship, while passing the nearby star, continues to move at constant velocity. They think their spaceship is still far from any star and undisturbed by gravity. Their experiments give unaltered results, and they continue to use special relativity as the natural means of explaining whatever happens.

We on Earth cannot feel the pull of the Sun. The Earth, in free fall, moves about the Sun always in such a way that its motion cancels the Sun's gravity. We feel the pull of the Earth's gravity, however, because we are surface-dwellers and not in free fall about the Earth itself.

The principle of equivalence asserts that the gravitational pull of all external bodies is locally annihilated in every free-falling system. The uncurved spacetime of special relativity that exists in the absence of gravity may, therefore, be used in free-falling systems. If there is indeed a grand theory of gravity, the principle of equivalence informs us that the grand theory must simplify to special relativity in free-falling systems.

□ □ □

The Newtonian universe failed to provide a grand theory of gravity. It had the serious defect that gravity acted instantaneously everywhere. When an apple fell to the ground, all places in the universe received the news simultaneously. Newtonian gravity ignored the speed limit of light and, in principle, could be made to do impossible things.

As an illustration, suppose that A (for Albert) and B (for Bertha) are in separate spaceships fleeing side by side from enemy X. In this science-fiction scenario of star wars it will not strain the credulity of the reader if we further suppose that all characters have Newtonian gravity guns that shoot at infinite speed.

Enemy X fires and destroys Albert. The act of firing and the act of destroying Albert occur simultaneously in enemy X's space. But Albert and Bertha have their own space and time, and for them these acts do not occur simultaneously. In fact, the act of firing by X occurs after the act of destroying Albert. On seeing the destruction of Albert, Bertha fires back immediately, and by destroying enemy X, she is able to save Albert.

An effect is thus cancelled after it has occurred by eliminating the cause. This violates the sacred law of causality: murder once done can never be undone. Causes cannot follow effects in such a way that effects can be undone after they have occurred.

Let us assume that a second enemy Y is stationary in the neighborhood of X. Our tale of space wars now takes a bewildering turn. Enemy Y sees the destruction of X and immediately fires back, and by destroying B saves X; A thereupon destroys Y and saves B; X then destroys A and saves Y; B now destroys X and saves A; and so on repeatedly, each erasing the cause of what has occurred. Faster-than-light transmission enables us to perform miracles and consequently is impossible in a rational physical universe.

Einstein was confident of the existence of a grand theory according to which gravity travels at finite speed, a grand theory that simplifies to special relativity in the local region of each system in free fall, a *general* theory applying to all observers everywhere, unlike the *special* theory applying only to observers in gravity-free regions.

□　□　□

Euclid in the third century B.C. at the Museum of Alexandria brought together the geometrical knowledge of the ancient world and established the Euclidean system of geometry. His geometrical system was not merely a Babylonian-Egyptian ragbag of rules but an analytical body of knowledge developed from a few explicitly stated assumptions. If you accepted the assumptions as reasonable and obvious, then step by step in logical order you were compelled to acknowledge the rest.

An important basic assumption of the Euclidean system of geometry is the parallel postulate. This states that through any point a straight line may be drawn equidistant from another straight line. All equidistant straight lines are parallel lines. In our bones we know that when two parallel straight lines extend to great distances they remain equidistant and never intersect each other. The parallel postulate cannot be shown to be true with absolute certainty, because all human experience occurs within a limited region of space. But it seems eminently sensible, and it was accepted without question by most geometers from the time of Euclid until the nineteenth century. A few geometers felt uneasy and tried to derive the parallel postulate from a more basic assumption. All attempts failed.

We now know that the parallel postulate is fundamental to Euclidean geometry. It enables us to distinguish the geometry of Euclidean space from the geometries of non-Euclidean spaces.

We find, when our observations are limited to what happens in small regions, that many non-Euclidean spaces have geometries closely resembling the geometry of Euclidean space. Small triangles and circles drawn in these spaces look much the same as our familiar triangles and circles. The simplest way to recognize a non-Euclidean space is to observe what happens to comparatively large geometric figures covering large regions of space. But this is not easy when large regions extend far beyond the range of normal human experience. Can we be fully confident that two straight lines, seemingly parallel to us in our local region of space, when extended beyond the limits of the largest telescopes, will stay parallel in distant regions?

Curved and uncurved two-dimensional surfaces may be visualized with moderate ease. Consider first a flat surface of unlimited extent. This is Flatland that possesses Euclidean geometry. Parallel straight lines drawn by two-dimensional Flatlanders in a small region, when extended to great distances, do not intersect in either direction. The parallel postulate holds true in Flatland. But equidistant straight lines drawn in curved surfaces, when extended to great distances, do not remain equidistant. Large triangles and circles in Curveland are not exactly the same as similar figures in Flatland. The uniformly curved surface of a sphere, for example, possesses non-Euclidean geometry.

A small region of a spherical surface is almost flat. The smaller a region in Sphereland, the flatter seems the surface in that region. If we were two-dimensional Spherelanders living in a very small region of Sphereland, we would believe that its geometry is Euclidean, and that Sphereland is actually Flatland. We might even find it impossible or exceedingly difficult to imagine Sphereland as not being Flatland. This is analogous to the situation in our world of three-dimensional space; we live in a comparatively small region, and we believe that space has Euclidean geometry. We find it impossible or exceedingly difficult to imagine what curved, three-dimensional space is like.

A straight line drawn on the surface of a sphere is a great circle. A Spherelander traveling in a straightforward direction follows a great circle and eventually returns to his starting point. He starts off in one direction and returns from the opposite direction. A great circle divides a spherical surface into two hemispheres. It is like a line of fixed longitude on the Earth that passes through both poles. Consider a second straight line, close to the first, also of fixed longitude. In a small region at the equator the two lines appear to be parallel. When extended, however, both lines intersect at the poles and therefore cannot be parallel. All straight lines on the surface of a sphere intersect one another, and the parallel postulate fails to apply. Three-dimensional space may have analogous spherical geometry; if we lived in such a space and always traveled in a straightforward direction, we would ultimately return to our starting point.

Passing through a point in flat space there exists one parallel, and only one, to any given straight line (this is the Euclidean postulate); through a point in spherical space of uniform curvature there exists no parallel to any given straight line; and through a point in hyperbolic space of uniform curvature (which I have not discussed) there exist many parallels to any given straight line. The parallel postulate of Euclid uniquely distinguishes Euclidean geometry and fails to apply to other spaces of either uniform or nonuniform curvature.

In our small part of the physical universe we think normally in terms of flat, three-dimensional Euclidean geometry and have utmost difficulty trying to visualize curved non-Euclidean space. Immanuel Kant went so far as to declare that non-Euclidean space is inconceivable and hence impossible. Euclidean geometry, he argued, is a priori (previous to experience) and "an inevitable necessity of thought." But he was wrong. On large scales the physical universe need not conform to ideas that derive from human experience on small scales. New experiences lead to novel ideas, and novel ideas lead to new experiences. Nowadays, we even talk of curved, four-dimensional spacetime.

Johann Gauss, a trailblazer in many fields of mathematics in the first half of the nineteenth century at the University of Göttingen, formulated the mathematical techniques for studying curved surfaces. If we were two-dimensional creatures living in a continuous and curved surface of any kind, unaware of a third dimension, we would survey and determine the geometry of our surface with the methods developed by Gauss.

But of the mathematicians who have contributed to our knowledge of non-Euclidean geometry, we remember most of all Gauss's brilliant young colleague, Bernhard Riemann. Riemann explored the metric properties of continuous spaces of two, three, and more dimensions and established the general equations defining their intrinsic curvature. Euclidean space is unique in having zero curvature; all other spaces have nonzero curvature that is either uniform (the same everywhere and in every direction as in Sphereland) or nonuniform (not the same everywhere and in every direction, as in Hillyland).

Riemann foresaw the possibility of an intimate relationship between geometry and physics. To most mathematicians and scientists his studies on the curvature of space seemed excessively abstract and divorced from reality. "Only the genius of Riemann, solitary and uncomprehended, had won its way by the middle of the last century to a new concept of science," said Einstein.

The mathematician William Clifford, who died when still a young man (Riemann also died when still relatively young), translated Riemann's work on geometry into English. In *The Common Sense of the Exact Sciences*, published posthumously in 1885, he championed the idea that geometry and physics are interconnected:

We may conceive our space to have everywhere a nearly uniform curvature, but that slight variations of curvature may occur from point to point, and themselves to vary with time. These variations of the curvature with time may produce effects which we not unnaturally attribute to physical causes independent of the geometry of our space. We might even go so far as to assign to this variation of the curvature of space "what really happens to that phenomenon which we term the motion of matter."

Clifford predicted the possibility of curvature waves (now referred to as gravity waves) and surmised, "this property of being curved or distorted is continually being passed on from one region of space to another after the manner of a wave."

A germinal idea was in the air. But special relativity had yet to be discovered, and the development of spacetime into a Riemannian world of dynamic curvature lay 30 years ahead.

Albert Einstein, born in Germany in 1879 (the year Clifford died), was an imaginative and inward-living person who did not respond to disciplined education. He was regarded as a backward child and an inattentive pupil. He read widely, thought deeply, developed an independent outlook on many subjects, and was mainly self-taught. Newton with his "silent face" and Einstein with his retiring manner, though having very different outlooks and personalities, shared much in common. Both had mystical religious and metaphysical views; both were not particularly distinguished as teachers ("there is too much education altogether," said Einstein); both were pestered by distracting adulation that made further scientific work difficult ("it is unfair and in bad taste," said Einstein). Each in his way had an extraordinary intuitive grasp of physical processes, and each pondered deeply for many years before producing his great theory of gravity.

Einstein's theory of general relativity reached final form in 1916, and to scientists and many sections of the public it seemed at last that Omar Khayyam's dream was fulfilled:

> *Ah love! could thou and I with him conspire*
> *To grasp this sorry scheme of things entire,*
> *Would not we shatter it to bits—and then*
> *Remould it nearer to the heart's desire!*

The Newtonian universe with its Euclidean geometry and inlay of mysterious gravity finally was shattered and remolded into a spacetime universe of varying geometrical curvature. The curved paths of the heavenly bodies, moving under the influence of gravity in the Newtonian universe of flat space, became the straight paths (or geodesics) in the Einstein universe of curved spacetime.

Einstein's equation of general relativity looks harmless enough,

$$R_{ij} - \tfrac{1}{2}g_{ij}R = \kappa T_{ij}$$

but actually is ten equations expressed in compact notation. Terms on the left side of the equation deal with the curvature of spacetime, and terms on the right deal mainly with matter and energy. Simple-mindedly we may think of geometry and matter as equated,

$$\text{geometrical strain} = \kappa \times \text{material stress}$$

in the sense that the strain or curvature of spacetime is proportional to the stress of matter or energy. Distant as well as local matter produce spacetime curvature. Matter influences the curvature of spacetime, and curvature of spacetime influences the motion of matter.

A flexible rubber sheet stretched flat illustrates what happens. The flat sheet represents uncurved space in the absence of gravity. Ball bearings when rolled on the flat surface follow uncurved trajectories. A heavy ball placed in the center of the sheet produces a depression, and the rubber sheet is then everywhere curved. This represents the curvature of space produced by a star and shows how matter affects local as well as distant curvature. The curvature near the ball is larger (stronger gravity) and far from the ball is smaller (weaker gravity); the curvature diminishes with distance, and at large distances the sheet is almost flat. Ball bearings when rolled on the curved surface follow curved trajectories similar to those of planets and comets under the influence of the Sun.

General relativity simplifies to Newtonian theory when gravity is weak (spacetime is almost flat) and bodies have speeds small compared with the speed of light. In the Solar System, where gravity is moderately weak and the planets move at comparatively low speeds, Einstein's master equation reduces to the Newtonian equations of motion and gravity. Some small discrepancies remain, however, and years of research have been devoted to the detection of these residual effects of general relativity in the Solar System. The results, in good agreement with predictions, have inspired confidence in the validity of Einstein's theory of gravity.

General relativity is not an easy theory. Or, rather, I should say that curved spacetime is not as easy to understand as Newtonian gravity.

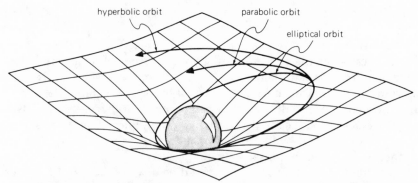

hyperbolic orbit   parabolic orbit

elliptical orbit

A stretched rubber sheet, depressed by a heavy ball, illustrates the deformation of space. (E. R. Harrison, *Cosmology: The Science of the Universe*, Cambridge University Press, New York 1981.)

Consider two stars in the Newtonian universe. The gravitational force that each exerts may be calculated as if the other did not exist. At any point in space the separate forces exerted by each star can be added to give the combined force exerted by both stars. But in general relativity this is not possible. The separate curvatures produced independently by each star cannot be simply added to give the combined curvature produced by both stars. Spacetime self-interacts in a most curious manner. The curvature produced by one star alters the curvature produced by the other. The following remarks might help to make clear this important aspect of general relativity.

You must take my word that all forms of energy have mass. Thus heat, which is a form of energy, has mass. Boiling water in a kettle weighs a billionth of a gram more than when cold. You must also take my word that spacetime curvature is, in effect, yet another form of energy. The curved spacetime around a star has distributed energy and therefore manifests its own effective mass.

The curved space (I mean, of course, curved spacetime) around a star, because of this associated energy and equivalent mass, acts as a source of additional gravity. Consequently, curved space is itself a source of further curvature. In the Newtonian universe gravity is not a source of gravity, but in the Einstein universe curvature is a source of curvature. The curvature produced by one star, because of its associated energy and equivalent mass, interacts with and modifies the curvature produced by the other star, and the combined curvature is not obtained by simply adding together the separate curvatures of the two stars. Self-interaction of space is the essence of general relativity.

Energy distributed in the curved space around the Sun produces more curvature and contributes to the Sun's distant gravitational pull on the planets. For this reason the Sun's gravity fails to obey precisely the inverse square law, and the planets do not follow exactly Kepler's elliptical orbits.

Consider two stars circling around each other. They move in curved orbits because of the curvature of space (or rather of spacetime). The orbiting stars cyclically strain and warp the surrounding space. This cyclic warping streams away as gravity waves at the speed of light. The waves transport energy, and the drain of energy from the binary system causes the two stars to spiral slowly toward each other. Such an effect has been observed and studied by Joseph Taylor at the University of Massachusetts in the case of a binary pulsar system.

Einstein's equation may be viewed as a wavelike description of gravity showing how curvature disturbances in the form of gravity waves travel at the speed of light. Gravity waves are not easy to observe because of the difficulty in extracting from them morsels of energy, and they have not been directly detected, despite many years of painstaking research. Emily Dickinson in *Time and Eternity* wrote,

> I never saw a moor,
> I never saw the sea;
> Yet I know how the heather looks,
> And what a wave must be.

We have never seen gravity waves, yet know what they must be, and we have little doubt they exist.

Curved space is suffused with energy and hence clothed in intangible mass. Had Aristotle and Descartes known, they would have hugged themselves with joy. The ancient atomists believed that nothing exists except atoms and the void, but Aristotle and later Descartes insisted that space could not exist by itself as an empty void and must be clothed in material raiment.

But actually, despite everything, the Newtonians have been vindicated. They believed space was reified by spiritual essence and therefore real in its own right. The nature of spacetime in general relativity theory seems more etheric in the Newtonian sense than material in the Cartesian sense.

Out in the depths of space something has happened in our free-falling windowless spaceship. The experimenters have emerged through an open hatch and are now observing the external world of cavorting bodies and whirling systems. Everything seen in the firmament acts in accord with the grand picture of general relativity. The free-falling experimenters in their own locally flat spacetime see the celestial bodies following curved paths as if under the influence of a mysterious, long-range astral force called gravity. But actually, all the freely falling bodies follow straight paths in curved spacetime.

□   □   □

Gravity normally is weak, as in the Solar System, and the Newtonian picture suffices for most of science and much of astronomy. But gravity may sometimes be strong and then produces astonishing effects. Bodies "having not the law," said Saint Paul on another subject, "are a law unto themselves." Black holes—monsters of the deep—having not the Newtonian law, are subject to the higher law of general relativity.

Newtonian gravity warns us that odd things happen when stars are either dense or massive. John Mitchell, rector of Thornhill in Yorkshire, pointed out in 1784 that a particle escaping from the Sun must move at a speed at least one-five hundredth the speed of light. He argued that if a star had the same average density as the Sun (slightly greater than the density of water), and a diameter more than five hundred times greater than the Sun's diameter, then not even light could escape from the surface of the star. "All light emitted from such a body would be made to return to it by its own power of gravity," he said. The star, though invisible, would still be detectable because of the effect of its strong gravity on the motions of satellites and nearby stars. The astronomer William Herschel was much intrigued by Mitchell's argument and thought that many luminous interstellar clouds could be interpreted as regions of trapped light.

Newtonian gravity foretells the possibility of these strange bodies that we nowadays refer to as black holes, and general relativity theory enables us to understand them. A black hole is born when a star collapses catastrophically to very high density.

Imagine a shrinking star. Gravity at its surface steadily increases and eventually becomes intense. Gravity is the old name for what is now known to be the curvature of space (or rather spacetime). As the star shrinks, the curvature of surrounding space increases, and finally space becomes sufficiently curved to encapsulate the shrinking star. It has become a black hole from which neither light nor anything else can escape. The Sun would become a black hole if its diameter decreased to 6 kilometers. The size of a black hole varies in proportion to its mass, and the Earth would have to shrink to the size of a golf ball to become encapsulated in its own space. The larger a black hole, the lower its density, and large black holes have low densities. Thus a black hole a billion times more massive than the Sun has a size roughly equal to the size of the Solar System and a density about the same as the density of air.

Inside a black hole everything collapses with utmost rapidity in a crescendo of rising density. But to a person outside at a safe distance nothing happens; the black hole seems frozen in a state of suspended animation, and at its surface time appears to stand still. (The deformation of spacetime means that time as well as space is altered.) Seen from inside everything falls catastrophically, and nemesis awaits only a moment away; seen from outside nothing changes because of the extreme deformation in spacetime.

We may think of a black hole as a region into which space flows from the outside world. Most black holes probably rotate, and we should, therefore, imagine the inflowing space as swirling like water draining away down a sink. To most of us, accustomed to thinking of space as little more than a vacuous nothing, this will seem an incredible way of visualizing a black hole. Remember, however, that space is no longer an intangible nothing. At the surface of a black hole—the event horizon—space flows inward at the speed of light.

Light rays outside the black hole travel through infalling space and may escape. At the surface, however, space falls at the speed of light, and rays seeking to escape through the infalling space remain stationary. The surface or event horizon of the black hole is the country of the Red Queen of *Alice in Wonderland*. "Now, *here*, you see," said the Red Queen to Alice, "it takes all the running *you* can do, to keep in the same place." Space inside the black hole falls faster than the speed of light, dragging everything with it, and nothing may escape, including light.

Far from the black hole spacetime is almost flat and practically the same as in special relativity. Near the black hole spacetime is greatly curved and very much deformed. As we approach the black hole, and the curvature increases, we see less and less of what lies ahead. At the last moment, before being engulfed, we see only what lies behind in the outside world and nothing of what lies ahead. Nature takes pity on us and veils from view the awaiting doom.

If by mischance we fall into a massive black hole of low density, we might quite easily not realize that anything unusual has happened; only slowly would it dawn on us that we are caught in the grip of a black hole, and an awful and unseen fate lies ahead.

□　□　□

Stars in their death throes are thought to be the birthplaces of black holes. When a star has consumed its central supply of hydrogen, it swells up and becomes a red giant.

The Sun in about 5 billion years time will also become a red giant. Its core or central region will contract, and at the same time its mantle or outer region will expand and fill the sky, engulfing Mercury and Venus, perhaps even the Earth. After some tens of millions of years as a red giant, the Sun will puff away its inflated mantle and reveal a condensed central core. It will then be a white dwarf, having a size about equal to that of the Earth and will bathe the dead terrestrial surface with a pale white light scarcely brighter than present moonlight.

Stars more massive than the Sun do not give up the game so easily. After becoming red giants they convert their helium into heavier elements, carbon, oxygen, and so on, all the way to iron, unlocking more and more nuclear energy. Some of this energy spills over and is used to manufacture elements heavier than iron, such as gold and uranium. The star has become fiercely bright, demanding more and more nuclear energy, and its reserves are soon exhausted. Only gravitational energy then remains, and to draw on this supply of energy the core must continue to shrink, getting denser and hotter. The core now generates neutrinos that stream out of the star, adding to the loss of energy.

The central density continues to rise and eventually reaches a point at which the heavy elements in the core are crushed and broken into helium. Soon thereafter the helium dissolves back into its constituent protons, neutrons, and electrons; the electrons are squeezed into the protons, leaving neutrons as the dominant survivors in the contracting core.

For tens and even hundreds of millions of years the bright star has radiated into space an immense amount of energy obtained by burning hydrogen into helium. Now, confronted with the dissolution of helium, it must repay all this energy instantly. At death's door, faced by ruinous debt, the star does the only thing possible: it draws on gravitational energy by collapsing catastrophically. The inrushing core terminates as a neutron star, and the outrushing mantle signals the birth of a new star. The mantle explodes, and a brilliant supernova briefly outshines all the other stars in the Galaxy.

A fraction of the heavy elements escapes into space and intermingles with the interstellar gas from which new stars and their planets form. Look at a coin and ask, Where was this metal made? It was made in the Promethean fires of a stellar nuclear reactor that died more than 5 billion years ago before the birth of the Sun.

From the death of the old star a neutron star is born. One thimbleful of neutron matter would weigh a billion tons on Earth. Quite likely, the newborn neutron star is a pulsar sending out a pulselike message of matter racked to its uttermost limits. Its searchlight beam sweeps across the sky, and we observe the beam periodically as it passes the Earth. At first the pulsar gyrates rapidly, perhaps hundreds of times a second; it turns slower and slower, and after millions of years sobs into silence.

Larger stars with more massive imploding cores cannot terminate as neutron stars. The neutron matter in inrushing cores more massive than about three times that of the Sun is unable to withstand the intense pull of gravity. These cores continue to collapse and become black holes. The laws of nature as we understand them lead to this conclusion, and it seems safe to say that many massive stars at the end of their evolution give birth to black holes.

William Herschel kept his eyes open for astronomical evidence of the existence of black holes. The search has picked up pace in recent years, and the mounting evidence now seems fairly convincing.

Many massive stars are members of binary systems. These stars orbit each other, sometimes in close embrace, exchanging matter and evolving in spectacular ways. When a massive member of a binary system collapses, it becomes either a neutron star or a black hole. Gas then flows from the companion, spiraling inwardly to the surface of the collapsed star, and some of the gravitational energy released is radiated away in the form of X-rays.

The study of these powerful X-ray sources indicates in a few cases that the collapsed companion is sufficiently massive to be nothing less than a black hole. Naturally, we are unable to see these black holes, but the radiation emitted by infalling gas betrays their presence; also, they affect strongly the motion of their companions, as the Reverend Mitchell foresaw.

□ □ □

A black hole voraciously consumes all that it encounters and is aptly described by the words of Jonathan Swift:

> All-devouring, all-destroying,
> Never finding full repast,
> Till I eat the world at last.

Once born it grows and puts on weight. The surrounding gas spirals in and is sucked up. Incautious stars straying too close are torn to shreds by tidal forces, and their wreckage adds to the headlong rush of "atoms and systems into ruin hurled." The inwardly spiraling gas, squeezed to high temperature, frantically radiates, and an appreciable fraction of the accreted mass is transformed into escaping radiation. A black hole on the prowl is an efficient engine that converts into radiation a fraction of whatever mass it devours.

The central regions of giant galaxies swarm with closely packed star systems. Consider what happens when black holes are unleashed among these rich star systems: they become star destroyers, devouring everything, even one another, never finding full repast. After hundreds of millions of years, a black hole attains a mass possibly a billion times that of the Sun. During its growth it pours forth a torrent of radiation, perhaps many times more intense than our entire Galaxy of stars.

Many astronomers are inclined to think that the distant and brilliant quasars are massive black holes accreting matter in the nuclei of giant galaxies. According to this theory a quasar becomes quiescent when a black hole has swallowed most of the surrounding gas and stars. Also, when a black hole has grown extremely large, plunging stars fall straight in without first being disrupted by tidal forces, and stellar wreckage ceases to contribute to the inward spiraling of luminous gas.

Possibly, quasars have a lifetime of a billion or so years, and the majority of them perhaps existed shortly after the birth of galaxies. Because they existed long ago, their light has taken billions of years to reach the Earth; consequently, most of the quasars seen by us lie at vast distances.

□ □ □

Einstein showed how mysterious gravity can be interpreted as curved spacetime. In the dynamic picture of general relativity, however, the electromagnetic forces remain much the same as before, lacking a similarly lucid geometric interpretation. For years Einstein endeavored to unify the electromagnetic and gravitational forces within a fully geometrized picture of the universe.

With Nathan Rosen, Einstein wrote in 1935, "In spite of its great success in various fields, the present theoretical physics is still far from being able to provide a unified foundation on which the theoretical treatment of all phenomena could be based." The underlying idea of general relativity, while applicable to matter on the large scale, has "hitherto been unable to account for the atomic structure of matter and for quantum effects." Einstein never succeeded in his quest for a more unified picture. What he had achieved, though nearer to the heart's desire, fell a long way short of what he himself desired.

To geometrize everything, including the weak and strong subatomic forces, now seems a well-nigh hopeless task. New conceptual schemes are in the making, of grand unified theories that combine the electromagnetic, weak, and strong forces into a single protean force, and of exotic supersymmetry theories that ambitiously combine grand unification and gravity under the rubric of supergravity.

No doubt in the future much will be understood that to us is still murky and perplexing. By then almost certainly we shall have uncovered new riddles. The unknown will loom as large as before, possibly more so, and the heart will yearn for revelation, which when found will lead inevitably to the discovery of fresh mystery. The more we know, the more aware we become of what we do not know.

11

The Cosmic Tide

We live in the Solar System on the planet Earth that revolves with other planets around a star called the Sun. Light from the Sun hurrying at great speed takes 500 seconds to reach the Earth and 5 hours to reach the outermost planet Pluto. The Earth that to us seems large is dwarfed by the Solar System with its far-flung planets.

Starlight from the nearest stars travels for years before reaching the Earth. If we imagine the Sun having the size of a grain of sand, on the same scale the nearby stars would be at a distance of 1 hour's drive on an interstate highway. Scattered out to enormous distances in all directions are 100 billion stars that constitute the whirlpool system called the Galaxy. Our Galaxy—a glittering carousel of stars, across which light takes 100,000 years to travel, and around which the Sun journeys once every 200 million years—compared with the Solar System seems gigantic beyond ordinary comprehension.

Much has been discovered about the Galaxy: its many kinds of stars, blue, yellow, and red giants, sunlike stars, binary stars, white dwarfs, and dense neutron stars; its spiral disk, seen by us as the Milky Way, consisting of young and old stars, where clouds of gas and obscuring dust give birth to new stars; its enveloping halo of old stars and globular clusters; and much remains to be discovered.

Newton's universe of uniformly distributed stars has become

Wright's universe of galaxies. Out beyond the Galaxy in the depths of space lie numerous galaxies of different kinds.

Light from the nearest galaxies takes about a million years to reach the Earth, and we see them as they were when early human beings gazed at the sky with wondering eyes. Many galaxies are midget systems of a few hundred million stars. But not all. The neighboring giant galaxy of Andromeda at a distance of 2 million light-years, festooned with out-lying midget systems, is much like our Galaxy. Scattered here and there we see giant ellipticals, also giant spirals similar to our Galaxy, and sometimes supergiant galaxies a hundred times more massive still. Among the giant galaxies can be found the powerful and enigmatic radio sources.

Through telescopes we see the majestic galaxies stretching away like celestial cities to unlimited distances. We look out into the depths of space and see galaxies as they were billions of years ago at a time before life arose on Earth. In their midst gleam the starlike quasars.

The galaxies, as Kant foresaw, cluster together to form even larger systems. Clusters of galaxies come in all sizes. Our Galaxy and its great companion galaxy of Andromeda are the dominant members of a swarm called the Local Group. Most clusters are comparatively small like our Local Group and contain tens of galaxies. But richly populated systems, such as the Coma and Perseus clusters, contain galaxies by the thousands, and often in their central regions blaze the supergiant galaxies.

The clusters congregate into sprawling superclusters, interspersed with immense voids of darkness where few clusters appear to exist.

Galaxies are the cradles of life. Who can doubt for a moment the existence of life out there in the galaxies? Life that in many cases is probably far more intelligent than on Earth. Even if we were so begrudging as to concede that life exists on only one planet to each galaxy, then in the colossal observable universe there would still be trillions of planets inhabited by living creatures.

The last shadows of mythical anthropocentrism slink away before the astronomical grandeur of the physical universe.

☐   ☐   ☐

The history of cosmology unfolds a growing conviction that human beings cannot occupy a position of central importance in the cosmic scheme.

The assault on the mythic universe by Hellenic science, followed by the Copernican and Darwinian Revolutions, dethroned the human species. The cosmic center—which was first the nation, then the Earth, the Sun, and finally the Galaxy—has vanished from the physical universe. From science, with the help of theology and philosophy, emerges an outlook expressed by the location principle.

The location principle states: *it is improbable that human beings have privileged location in the physical universe.* Other planets encircle other stars in other galaxies in other clusters and may have life that in many instances is more advanced and even more precious than on Earth. Why, then, should human beings be singled out for special dispensation? The location principle asserts that of all the planets, stars, galaxies, and clusters of galaxies in the universe, it is improbable that the Earth, Sun, Galaxy, and Local Group are in any way uniquely privileged. We can be kings of the cosmic castle in the mythic universe but not in the modern physical universe.

☐ ☐ ☐

Seen from a ship, the sea stretches away the same in all directions. The waves that disturb the surface are no more than mere incidental irregularities. Lucretius echoed the atomists when he said, "the universe stretches away just the same in all directions without limit." Yet this ideal in the minds of the atomists is not in the least obvious to an observer looking out from Earth.

We are afloat, it seems, in a cosmic ocean surrounded by great waves that at first glance do not appear in the least like mere incidental irregularities. Only when we look out far beyond the Galaxy do we find that on the average the things seen in one direction look much the same as those in other directions. All directions look alike, and from our particular viewpoint the universe is therefore isotropic.

Discussions in modern cosmology place considerable weight on what is called the cosmological principle. The cosmological principle, so-named by the astrophysicist Edward Milne in 1933, was expressed by him in the words: "Not only the laws of nature, but also the events occurring in nature, the world itself, must appear the same to all observers, wherever they may be." The principle states that when local irregularities are ignored, or averaged out, the universe at any instant in time is the same everywhere in space. As Einstein said in 1931, "all places in the universe are alike." In other words, the cosmological principle asserts that the universe is homogeneous.

The concept of cosmic homogeneity originated with Anaxagoras and the atomists and reemerged in the late Middle Ages. Cardinal Nicholas of Cusa in the fifteenth century, using the analogy that God is ubiquitous and uncircumscribed, declared "the fabric of the world has its center everywhere and its circumference nowhere." What is potential in God must be actual in the created universe.

Nowadays astronomers observe isotropy (all directions are alike), and cosmologists postulate homogeneity (all places are alike). The probability argument of the location principle links together observed isotropy and postulated homogeneity.

Let us imagine that we stand at the summit of a hill from which the landscape around us looks much the same in all directions. The scenery from our vantage point appears isotropic. But we are not at liberty to declare that all places are alike and the landscape is homogeneous. When seen from any other point, the scenery is not the same in all directions because of the presence of the hill. Our isotropic view from the summit is the consequence of special location.

Similarly, if we occupy an assumed center of the universe, as in the Aristotelian and medieval universes, our special location explains why the fixed stars in all directions appear much the same.

Unfortunately, we are confined to a small region of the universe and cannot travel elsewhere to another vantage point—say a few thousand million light years away—to take a fresh look at the cosmic scenery. Instead, we must use the location principle that assures us that special location is improbable. Because of this principle we feel compelled to draw the conclusion that all directions are alike, not only from our particular region of the universe, but from all other regions. Isotropy probably is common everywhere and not unique to our region.

It comes to this: when irregularities are ignored, the observation that the universe is isotropic from our place in space, coupled with the location principle that says special location is improbable, leads to the conclusion that the universe probably is isotropic from every place and consequently is homogeneous.

We must imagine that we stand not at the summit of a hill, but on a flat or spherical surface, and all directions are alike everywhere in space simply because the surface is everywhere the same. Such a surface is homogeneous. Analogously, the universe is homogeneous, as asserted by the cosmological principle.

Apart from astronomical irregularities, cosmic space is either flat or uniformly curved. With cosmic space comes cosmic time ticking away everywhere at the same rate. If we could rush around the universe at infinite speed, we would find at any instant that everywhere seems much the same, the laws of nature the same, clocks running in synchronism, and things evolving in similar ways.

The cosmological principle—founded on astronomical observations and a probability argument—unites the universe into a homogeneous whole.

□   □   □

The expanding universe ranks among the most startling discoveries made in the twentieth century. The galaxies are drifting apart, the yawning gulfs of space are widening.

From the way the galaxies move apart we can estimate that long ago, somewhere between 10 and 20 billion years, everything existed in a state of extreme congestion commonly referred to as the big bang.

In recent years a second startling discovery has been made. The afterglow of the big bang suffuses the whole of space. This ubiquitous glow, invisible to the unaided eye, is the three-degree cosmic radiation discovered by Arno Penzias and Robert Wilson in 1965. Its rays travel freely in space, coming to us in all directions, and the extraordinary isotropy of this radiation reinforces our belief in the basic homogeneity of the universe. Although of very low temperature (3 degrees above an absolute zero that is 273 degrees below the freezing point of water), this radiation acts as an important constituent of the universe. Hold up the palm of your hand to the sky, day or night, and a thousand trillion photons—particles of light—of the cosmic radiation will fall on it in 1 second. This radiation, now cooled and enfeebled by expansion, long ago was the incandescent light of the early universe.

The Stoics believed in a universe of periodic fiery explosions and implosions. Edgar Allan Poe in his imaginative essay *Eureka* of 1848 vividly portrayed the possibility of a pulsating universe expanding and collapsing: "Are we not, indeed, more than justified in entertaining the belief—let us say, rather, in indulging a hope—that the processes we have ventured to contemplate will be renewed forever, and forever, and forever; a novel universe swelling into existence, and then subsiding into nothingness, at every throb of the Heart Divine." The idea of a cyclic universe expanding and collapsing, forever bouncing from big bang to big bang, each a throb of the heart divine, each a day in the life of Brahma, still persists to this day.

Edwin Hubble's observations in the late 1920s and early 1930s, with contributions by other astronomers, have made secure the idea of an expanding universe. Often it is too troublesome to trace an idea or a discovery back to its origin and is more convenient to make attribution to the one who convinced the world. As the New England poet James Lowell said,

> *Though old the thought and oft expresst,*
> *'Tis his at last who says it best.*

Rightly or wrongly, we attribute to Hubble the discovery of the expansion of the universe.

□   □   □

The story begins in 1912 with Vesto Slipher of the Lowell Observatory measuring the shift in the spectral lines of the light emitted by other galaxies. From his measurements of spectral shifts he calculated the velocities at which these galaxies were approaching and receding. By 1923, as a result of Slipher's painstaking work, it was known that of the forty-one galaxies studied, five were approaching and thirty-six receding. Clearly, if galaxies moved randomly in all directions, about half should be approaching and half receding. Slipher's observations showed that the galaxies had a mysterious tendency to recede.

On the front page of *The New York Times* in 1921, under the heading *Celestial Speed Champion*, Slipher reported his latest discovery of a receding galaxy:

> The lines in its spectrum are greatly shifted, showing that the nebula is flying away from our region of space with a marvelous velocity of 1100 miles per second. . . . If the above swiftly moving nebula be assumed to have left the region of the sun at the beginning of the earth, it is easily computed, assuming the geologist's recent estimate of the earth's age, that the nebula now must be many millions of light years distant. The velocity of this nebula . . . further swells the dimensions of the known universe.

The distant galaxies (or nebulae) are running away and in the past must therefore have been much closer together.

Albert Einstein and the Dutch astronomer Willem de Sitter proposed in 1917 quite different models of the physical universe based on the theory of general relativity. Einstein's version had spherical geometry and contained matter; it was closed, and a person traveling in a straight line for a long period of time would eventually return to the starting point from the opposite direction; moreover, it was static, neither expanding nor collapsing. The de Sitter version contained no matter; it was open, and space extended to infinite distance in all directions and could not be circumnavigated.

The de Sitter universe, with its pathological absence of matter, might have been ignored and forgotten but for one particularly interesting feature. Particles of matter when sprinkled in it shared a tendency to move apart. Cosmologists conjectured that this *de Sitter effect* might have some bearing on the results obtained by Slipher. A few years later cosmologists realized that the de Sitter universe was expanding. An apt distinction was then drawn: the Einstein universe consisted of "matter without motion," and the de Sitter universe consisted of "motion without matter."

The German astronomer Carl Wurtz, prompted by Slipher's velocity measurements, proposed in 1922 a sort of velocity-distance law, according to which the further away a galaxy, the faster it recedes. He estimated the distances of galaxies by their apparent sizes and made the not very reliable assumption that the smaller the apparent size of a galaxy the greater its distance.

Also in 1922 the Russian physicist Alexander Friedmann published a paper *On the Curvature of Space* in a distinguished German scientific journal. In this pioneer work he investigated "nonstationary worlds" that either continually expand or first expand and later collapse. A second paper followed in 1924 in which Friedmann looked at alternate nonstatic worlds. Both papers drew little attention. Georges Lemaître, a Belgian mathematician and an ordained priest, published in 1927 a paper entitled *A Homogeneous Universe . . . Accounting for the Radial Velocity of Extra-Galactic Nebulae* in which he explored many of the characteristics of expanding universes. This work also drew little attention until translated into English 4 years later in the *Monthly Notices* of the Royal Astronomical Society of Great Britain.

Hubble meanwhile had undertaken the task of measuring the distances of nearby galaxies, using special pulsating stars of known intrinsic brightness, called cepheids. In 1928 Howard Robertson used Slipher's velocity measurements and Hubble's distance estimates to derive a linear velocity-distance law. This law was firmly established in 1929 by Hubble, and the available information showed clearly that the galaxies are receding and the universe is expanding. Improvements in the art of estimating astronomical distances and extensive observations have since provided a clearer and more accurate picture of the expanding universe.

In his book *The Expanding Universe* Arthur Eddington wrote in 1933: "The unanimity with which the galaxies are running away looks as though they had a pointed aversion to us. We wonder why we should be shunned as though our system were a plague spot in the universe. But this is too hasty an inference and there is really no reason to think that the animus is especially directed against our galaxy." To youths such as myself Eddington explained that the galaxies are not running away from us but from one another. Astronomers in all galaxies have at first the impression that their particular galaxy is the plague spot of the universe.

☐　　☐　　☐

We come now to an important point. The universe does not expand against a background of passive space, and the galaxies do not hurtle away through a vacuous nothing. Such a view belongs to the archives of the Newtonian universe. In the twentieth century we have general relativity, in which space has become an active participant on the cosmic stage.

The universe consists of expanding space. It is not immersed in passive space but on the contrary contains dynamic space. Galaxies float at rest and are carried apart by the expansion of space.

Of course, the galaxies are never exactly at rest in expanding space. They have their local and random motions, usually within clusters, and this explains why some of the nearest galaxies are approaching and not receding. Also, most clusters do not expand, and only the space between them expands. We are interested in the main outline, however, and for simplicity shall think of the galaxies as unclustered and at rest in expanding space.

To make the picture clearer, let us consider an expanding sheet of rubber. We imagine that this flat surface represents the space of our physical universe.

The sheet expands uniformly. By this I mean it expands isotropically (the same in all directions) and homogeneously (the same at all places). A triangle drawn anywhere on the surface remains a similar triangle while its size steadily increases.

We draw a circle and declare that it represents a galaxy. But as the surface expands, this "galaxy" gets bigger. A real galaxy, held together by its own gravity, is not free to expand with the universe. If the circle is labeled star, or planet, or atom, this also would be wrong, because all such objects are held together tightly and are not free to partake in the cosmic dilation. We observe the expansion of the universe because our observatories and measuring instruments have fixed sizes.

We erase the circle and replace it with a small paper disk. As the surface expands, the disk remains constant in size, and hence we have found a way of representing a galaxy in an expanding universe.

Over the surface, more or less uniformly, we scatter a large number of paper disks of various sizes. They remain at rest on the expanding surface and retain their fixed sizes. From each disk the surrounding disks recede. Imaginary inhabitants on any one disk have the impression that they occupy the cosmic center from which everything is running away. But the inhabitants on all disks share this impression, and there is no actual center.

It is easy to demonstrate the velocity-distance law. We choose any disk and label it A. A second disk, labeled B, at a certain distance moves away at a certain velocity. A third disk, labeled C, in the same direction as B and at twice the distance moves away from A at twice the velocity of B. Obviously, by virtue of homogeneity, C must move away from B at the velocity that B moves away from A. The easiest way to demonstrate this is with a length of elastic that has attached paper clips spaced at equal intervals. As the elastic is stretched we see the paper clips move away from one another at relative velocities proportional to their spacings. The physical universe behaves in much the same way, and the farther apart the galaxies, the faster they recede from one another.

☐   ☐   ☐

The velocity-distance law seems simple enough to us looking down on the expanding sheet. Like gods surveying the Trojan Plain, we see what happens everywhere in cosmic space at every instant of cosmic time. But the inhabitants on the disks find the situation far from simple, and the terms velocity and distance used in the velocity-distance law need careful interpretation.

The disks rest on the surface and move apart because the surface expands. Relative to one another they have recession velocities. The galaxies at rest in expanding space also have recession velocities relative to one another.

But galaxies move to and fro inside their clusters and have peculiar motion in addition to recession motion. They have *peculiar velocity* as well as *recession velocity*. Peculiar motion—of local importance but not of great cosmological significance—applies to bodies that move through space. This peculiar or ordinary motion, familiar to us on the Earth, in the Solar System, and in the Galaxy, when measured locally, is subject to the rules of special relativity. Peculiar velocity never exceeds the velocity of light.

Recession velocity applies to motion produced by the expansion of space and is exempt from the rules of special relativity. Failure to distinguish between peculiar velocity and recession velocity leads to endless confusion for the beginner in modern cosmology. Not the peculiar velocity of bodies moving through space, which is familiar, but the recession velocity of bodies moving with space, which is unfamiliar, must be used in the velocity-distance law.

The measurement of distance with a tape measure is no great problem for us who gaze down on the expanding surface. But from our god-like eminence, even we must exercise a little care and ensure that distances in the velocity-distance law are measured at the same instant in time. When a disk at a distance of 1 meter from a chosen point recedes at 1 centimeter a second, we note that another disk at a distance of 10 meters recedes at 10 centimeters a second. Both distances are measured simultaneously. We cannot afford to delay until the second disk reaches a distance of 11 meters, for its recession then might not be exactly 11 centimeters a second. All distances must be determined at the same instant, because the surface may not be expanding at constant rate. The expansion may be speeding up or slowing down. Similarly, the universe may not be expanding at constant rate.

Measuring the distances of remote galaxies is an arduous under-taking. First we must know the distance to the Sun; then by means of parallax measurements we must find the distances of nearby stars. By comparing stars of known brightness we find the distance of star clusters farther away, and by comparing star clusters we reach out to greater distances. The bright cepheid variable stars acquire in this way known intrinsic brightnesses and now play an important role as indicators of distance. Patient work with careful interpretation determines the distances of star clusters and cepheid variables in the nearest galaxies. For galaxies farther away we use whatever suitable distance indicators are available, such as the brightest stars, luminous clouds of gas, and supernovas. Very luminous galaxies of known intrinsic brightness then become beacons that help to determine the distances of rich clusters, some so distant that small groups like the Local Group are undetectable.

We see the galaxies as they were in the past long ago, and allowance must be made for their evolution and change in brightness. The whole subject of distance measurements in astronomy is a highly intricate and precarious art in which uncertainties mount up and errors unavoidably accumulate with distance.

We see the galaxies not where they are now but where they were when the light we see was emitted. The estimated distances must be adjusted to a common instant of cosmic time, say the present epoch, before using them in the velocity-distance law. Estimating the distances of galaxies is difficult enough, but adjusting these distances to a common epoch makes the problem even more difficult. The adjustment requires that we know not only how the galaxies evolve, but also how the expansion of the universe changes with time.

We still lack precise knowledge of how fast the universe expands. Also, we are not sure how the expansion changes with time. Cosmologists feel confident that the expansion is slowing down, yet the figures quoted for the observed deceleration remain very uncertain and must be taken with a grain of salt.

For every million light-years of extragalactic distance the recession velocity increases by about 20 kilometers a second. This expansion rate is ten times less than that first estimated by Hubble with his limited information and rough-hewn estimates of distance, and it is still uncertain by perhaps a factor of two.

Our rubber-sheet model of the expanding universe has helped us to understand the simple velocity-distance law. (Not simple for observers but simple for us looking down on the expanding surface.) Before leaving the model we perform one more experiment.

We arrange first that the expansion occurs at a constant rate. Then on the expanding surface we constantly sprinkle fresh disks in such a way that nothing ever seems to change. New disks occupy the widening gaps between old disks, and the average separating distance between the disks remains unchanged. This illustrates what happens in the steady-state universe proposed in 1948 by Hermann Bondi, Thomas Gold, and Fred Hoyle: new galaxies form continually from freshly created matter, and the cosmic scenery remains eternally unchanged. All places on the average are alike in time as well as in space.

We know that for every million light-years of distance the recession velocity increases by about 20 kilometers a second. If we divide 1 million light years by 20 kilometers a second, we get 15 billion years, which is a crude estimate of the age of the universe. In the 1920s and 1930s the recession velocity was thought to be ten times larger, thus giving an estimated age of from 1 to 2 billion years. This puzzling result, implying that the universe was younger than the Earth, created a paradox that lasted until the 1950s. The steady-state universe, without a beginning and an age limit, was an ingenious way of circumventing the paradox.

The steady-state theory provoked considerable controversy, echoing the heated debates between the catastrophists (big-bangers) and the uniformitarians (steady-staters) in the early nineteenth century. A self-replicating universe of continuous creation, eternally unchanging in appearance, though fascinating to some persons was repugnant to others.

More precise measurements have since increased the estimated age of the universe, and it is now reckoned to be sufficient to accommodate the oldest stars. The original main purpose of the steady-state universe no longer exists.

The three-degree cosmic radiation discovered in 1965 indicates that the universe probably was once exceedingly dense and hot, and it seems safe to say that the physical universe cannot be eternally unchanging in appearance, as supposed by the proposers of the remarkable steady-state theory.

□　□　□

The recession velocity, according to the velocity-distance law, increases steadily with increasing distance. Eventually we reach the edge of the Hubble sphere. At a distance of roughly 15 billion light-years—the radius of the Hubble sphere—the recession velocity matches the velocity of light. Inside the Hubble sphere things recede slower than the velocity of light; outside the Hubble sphere things recede faster than the velocity of light. How is it possible for anything outside the Hubble sphere to move away faster than light?

We must realize that the universe has no edge and cannot terminate abruptly. Either space extends to infinity or curves back on itself like the surface of a globe. Homogeneity in both cases requires that the recession velocity progressively increases and eventually exceeds the speed limit of light outside the Hubble sphere.

There are as many Hubble spheres as galaxies. Each galaxy has its own Hubble sphere, at the surface of which the recession velocity from the galaxy matches the velocity of light. The cosmic edge cannot exist at the surface of our Hubble sphere, because all galaxies have Hubble spheres, and the universe would have as many cosmic edges as galaxies.

Suppose for a moment that the surface of our Hubble sphere were indeed the cosmic edge. Nothing now exists outside, and hence nothing recedes faster than the velocity of light. Those who want to save the universe from the unpalatable thought of distant objects moving away faster than light may breathe a sigh of relief. But at what a cost! Other galaxies are denied similar Hubble spheres and condemned to a lopsided view of the universe. All places are no longer alike. We have restored our privileged position at the center of the universe and, by throwing away the location principle, have lost assurance that things out there obey laws similar to those here. We might as well give up cosmology as a subject of scientific inquiry.

If we hold to the location principle, as modesty suggests, then all galaxies stand on equal footing and have their own Hubble spheres. Inevitably, this means that things exist outside our Hubble sphere and recede from us faster than the velocity of light.

How can this be possible? The answer is because the universe expands not in space but consists of expanding space. Special relativity with its speed limit works locally but not globally.

□ □ □

Light travels at constant speed measured locally in the space through which it travels. Nothing in nature has an ordinary speed exceeding the speed limit of light. All ordinary or peculiar velocities are subject to this limit, but recession velocities are without limit.

The velocity-distance law tells us that the greater the distance, the greater the recession velocity. At infinite distance in an open universe the recession velocity is infinitely great. Recession occurs because of the expansion of space and does not consist of ordinary motion through space. Those persons who find it difficult to understand how recession can be without limit generally make the mistake of supposing that the galaxies are shooting away through space like projectiles; they have not been told that the galaxies are at rest in expanding space.

Consider a galaxy outside our Hubble sphere. Light rays from the galaxy, emitted in our direction, hurry toward us and travel through space that recedes faster than the speed of light. Thus even the light emitted by the galaxy recedes from us. As Arthur Eddington said in *The Expanding Universe*, "light is like a runner on an expanding track with the winning-post receding faster than he can run."

A galaxy at the edge of the Hubble sphere recedes at the speed of light. Its rays emitted in our direction stand still relative to us. Here again is the country of the Red Queen, where however fast Alice runs she stands still and gets nowhere.

Outside the Hubble sphere even light is receding. It would be a mistake, however, to suppose that our Galaxy will never receive this light. The Hubble sphere itself expands, generally faster than the universe, and its surface sweeps out and overtakes the distant galaxies. The inflating Hubble sphere contains more and more of the universe, and the number of galaxies inside increases by about ten each year. A galaxy outside the Hubble sphere may one day be overtaken; it will then lie inside, and its emitted light rays at last will be able to approach our Galaxy and be received. Eddington's runner must not give up the race, but keep on running, because the expanding track may be slowing down, and the winning post will one day be reached.

We cannot see the whole universe, only that part around us referred to as the observable universe. There are a few technical complications concerning the horizon of the observable universe that need not bother us in this discussion. The observable universe inflates in much the same way as the Hubble sphere, and in the course of time we see more and more of the distant universe. As a rough-and-ready guide we may assume that the observable universe is the Hubble sphere.

☐   ☐   ☐

We know that the universe expands because of the redshifted light received from distant galaxies. Red light has longer wavelengths than blue light. The blue light emitted long ago by a distant galaxy is received by us as red light. The light is not reddened by the removal of blue light, as in a London fog, but all wavelengths are redshifted. This displacement—or redshift—toward the red end of the spectrum is a consequence of the expansion of the universe.

Light rays journey for long periods of time, over vast regions of expanding space, and what happens is quite simple and easy to understand. The rays are stretched by the expanding space through which they travel. The wavelengths expand, and therefore, blue light slowly changes into red light. Take a length of stiff wire and bend it into a wavy

A wave of radiation stretches as it travels in expanding space.

or snakelike shape; now slowly pull on both ends and notice how the waves get longer. This is analagous to what happens to waves of light traveling through expanding space. The light received is redder than the light emitted, and the more that space expands while light travels to us, the greater the redshift.

A galaxy emits rays of light that eventually are observed in another galaxy far away. While the rays travel through expanding space between the two galaxies, their wavelengths steadily increase. Finally, the rays enter a telescope in the receiving galaxy, and the astronomers notice that the wavelengths have increased by a certain amount. They study the spectrum, comparing it with the spectra of luminous sources in their own galaxy, and in this way determine the amount of the redshift. The astronomers assume that the light-emitting atoms in their own galaxy are similar to the light-emitting atoms in the distant galaxy. They, in fact, assume that the universe is homogeneous in the sense that atoms and the laws of nature are everywhere the same. To justify this far-reaching assumption, they observe that the galaxies in all directions appear much alike, and by invoking the location principle, they deduce homogeneity.

Let us suppose the astronomers discover that the received rays have their wavelengths increased twofold. From this amount of redshift they know immediately that the universe has expanded twofold since the rays were emitted. During that time the average density of matter in the universe has decreased eightfold.

Expansion redshifts are extremely useful, but not for the reason most persons think. These redshifts measure directly the expansion of the universe during the time elapsed between emission and reception. They do not tell astronomers how fast the galaxies recede, or how far away they are, or how long ago they emitted the light now received; all this information must be deduced within the framework of a theoretical model. Instead, they tell astronomers precisely how much the universe has expanded between the emission and reception of light.

Let us suppose that a distant galaxy emits a pulse of light once every second, and we in our Galaxy detect these pulses of light. The pulses on leaving the emitting galaxy have initially a separation in space of 1 light-second. While they travel through the intervening and expanding space their separation increases. The pulses, therefore, are not received by us at a rate of one every second, but at a slower rate, because their separation in space is now more than 1 light-second. When the redshift informs us that the wavelengths have been increased twofold, the pulses arrive with a separation of 2 light-seconds, and we receive them at 2-second intervals. If the pulses are emitted once every year as measured by a clock in the emitting galaxy, they are received once every 2 years as measured by an identical clock in our Galaxy. A person in the distant galaxy living for three score and ten years appears to live for seven score years.

At large redshifts everything seems to change much more slowly than locally. Out near the horizon of the observable universe, where all appears extremely redshifted, time has slowed down to a snail's pace, or so it seems to us here.

□   □   □

In some popular treatments of modern astronomy and cosmology the reader gains the impression that the physical universe is a world of extreme violence, where the galaxies shoot away through space like the ejecta of an explosion and are themselves the scenes of dreadful cataclysms.

If we must be anthropocentric, then the people of the high Middle Ages, with their universe of harmonious spheres, were perhaps closer to the truth than the present portrayers of cosmic violence. The stately orbs and their measured tread in unison to the music of the spheres have in the modern universe become the galactic cities, alit with starlight, drifting serenely in the cosmic tide. The music of the spheres has become the melody of space and time and their harmony the symphony of wondrous forces. Once, long ago, it all began in an age of brilliant light.

The closer we examine the design of the majestic physical universe, the more we marvel at its harmony and beauty and the more we appreciate its fitness for habitation by life.

12

Do Dreams Ever Come True?

Natural historians would love to search the past in a Wellsian time machine and return to tell the "tales of long, long ago, long, long ago" that in the words of Thomas Bayly, a nineteenth century ballad writer, "to us are so dear." These historians little know that a timeship has been invented by a professor in the Department of Fantasy and Witchcraft at the University of Massachusetts. In this diachronic conveyance we shall take a journey—a safari in time—back to earlier periods of cosmic history.

Let me welcome you aboard with these comments. Moving backward through time is an uncommon way of presenting history, and to avoid the incongruities of a movie shown in reverse, I shall occasionally stop the machine and allow time to resume its normal Newtonian flow while we gaze at the scenery. Incidentally, I must warn you that our timeship is still in an experimental stage and will not always do exactly what we want. Please fasten your seat belts.

Tentatively I start the timeship in reverse gear, and it lurches into motion. Its dials spin alarmingly, and though I slam on the brakes almost immediately, we have already traveled 2 million years. Through the windows we see hominids striding around in the early Pleistocene. Uncontestably it would be very interesting to stay and observe their progress. But the controls of the new machine seem a bit stiff, and I have yet to learn how to make the nice adjustments necessary for time steps of less than 1 to 2 million years.

We move on. At 10 million years we again stop briefly and see the hominoids converging into the primate ancestry of apes and humans. Nothing in the sky has changed—the stars and galaxies look much the same—yet we cannot help noticing how the continents are adrift on the Earth's surface. At 20 million years—with another brief pause—primates first appear, and about this time India slams into Asia and thrusts up the Himalayan ramparts.

On we go, across the Cenozoic* era, skipping back some 65 million years. We arrive in time to witness the ascent of the mammals, the flourishing of grasses and flowering plants, but we have unfortunately just missed the demise of the dinosaurs. We cross the Mesozoic era, pausing in the Jurassic period to photograph a few large dinosaurs, and after a journey of 200 million years we reach the Paleozoic era in time for lunch. The land masses have temporarily fused together, forming the supercontinents of Laurasia to the north and Gondwanaland to the south, separated by the Tethys Sea.

After a stroll through the exotic forests and a glance around at the enormous inland seas and marshes teeming with amphibia, fish, and insects, we climb aboard and recommence our journey. With its dials spinning (and digital displays jumping) the timeship leaps across the Paleozoic era to a time 600 million years before the present. The great forests and all the reptiles, fish, and insects have gone, leaving a silent world to metazoans and other invertebrates.

We enter the long Proterozoic era. An unrecognizable Earth, no longer populated with multicellular creatures, swarms with unicellular forms of life. Overhead the Sun pours down its unfiltered ultraviolet rays.

Of course, if the news leaked out about our timeship, natural historians would be flocking to Massachusetts, queuing up to find the answers to all sorts of questions and to solve all sorts of problems: to find the answer to why species evolve spasmodically and to solve the problem of periodic mass extinctions of dominant life forms; to discover, among other things, why life in general transforms rather suddenly at the end of the Proterozoic era and whether the ozone layer in the upper atmosphere, shielding the Earth's surface from the Sun's ultraviolet rays, has something to do with cohorts of cells uniting into multicellular creatures. As yet the news has not leaked out.

Our journey backward in time must now proceed at a less-leisurely rate. While the eons roll by we notice the distant galaxies creeping closer. Constellations of newborn stars twinkle in the night sky, and multitudes

---

*Ceno means recent, meso middle, paleo ancient, and protero earliest.

of dying stars flare up and extinguish. Conceivably, life originates in myriads of solar systems, and perhaps in most it never gets very far.

After a total journey of 5 billion years, at a time when the Galaxy is roughly half its present age, we stop for dinner and a night's rest. The evening's entertainment consists of watching the birth of the Solar System.

Far from the center of the Galaxy, in a large interstellar cloud, lurks a smaller, denser, darker, cooler cloud of churning gas and dust that slowly contracts. As the small, dark cloud shrinks it swirls more rapidly. After millions of years its center grows dense and hot and develops into our embryonic Sun. Meanwhile, colliding grains of dust coagulate, and meteoroidal rock and ice settle into an encircling disk. Planetesimal bodies grow by slow accretion and develop into the planets and their satellites. The primordial Sun brightens, and in a flurry of convulsive vigor, thrusts back into space the remaining uncondensed gas of the dark cloud.

The Sun and its orbiting planets with their attendant satellites then sweep up much of the flotsam and jetsam littering interplanetary space. Evidence of this spectacular period of bombardment, lasting hundreds of millions of years, is still visible on the cratered face of the Moon, and on the surfaces of planets, such as Mercury and Mars.

Here is an opportunity that cannot be missed. We land on Earth, and with the timeship set in forward gear, we probe the future in quest of the origin of terrestrial life.

The Earth spins rapidly. Through veils of dust we see the Sun careering across the sky, with sunset following sunrise after only 2 hours. The glowering sky reflects the ruddy glare of lava flows, and the primitive atmosphere, fed by fiery volcanic plumes and tormented by monster storms, consists of nitrogen, ammonia, methane, water vapor, carbon monoxide, and choking gases, with only a trace of oxygen. The smoking and steaming surface, stalked by berserk giant tornadoes, heaves incessantly with earthquakes and the impact of large meteorites.

Blinding rays of sunlight and constant frenetic lightning conspire to establish a worldwide biochemical industry, which in the course of time manufactures immense quantities of an array of organic compounds, including a miscellany of amino acids and nucleotides.

High mountain ranges and deep oceans are landscape features of the future. But shallow seas abound, each a laboratory pool of primeval broth, continually boiled, shaken, and decanted. Countless myriads of biochemical experiments are performed in the seas and atmosphere every second. After hundreds of millions of years the nucleotides form into chainlike molecules of various codings, affecting the assembly of amino acids into proteins of numerous kinds.

At some stage the inevitable happens: a quixotic molecule becomes self-replicating and begets a protean species whose basic design endures beyond the lifespan of its individual members. We cannot see clearly through the fog and torrential rain, and we must infer that an entire sea becomes dominated by a dynasty of replicating molecules. Possibly, and here the gloom seems thicker, many seas discover their own species of replicating molecules. The living seas compete, exchanging their genetic codings by inundations, interconnecting streams, and windborne foam.

Half a billion years later the cells emerge in simplest form. The dilution of the seas, with water brought from the Earth's interior by volcanoes, encourages the development of replicating molecular systems capable of retaining their own rich environment of organic compounds. The invention of membranes, enclosing tiny autonomous organic worlds, ushers in the cells.

Cells are the smallest but possibly not the first forms of life. The honor of being the first goes conceivably to the *thalassabionts*—if I may coin a word—that are the living seas in which originate the replicating molecules. Who knows whether the thalassabionts succeed in creating coordinated, sea-wide, membraneless structure? Who knows what subsequently happens to these strange creatures that we find difficult to recognize as living? Here are the ingredients of a story to outrival all science fiction.

The cells thrive, enfolding one another and forming composite organisms; by trial and error they evolve into miracles of intricacy, developing sexual and asexual cell division. Constant experimentation, supervised by natural selection, occurs throughout the 4 billion years of the Proterozoic era.

□   □   □

We have detoured much too far. Reluctantly, we reverse our machine and leap back to a period before the formation of the Solar System.

After a night's rest we embark once more on our journey. While the dials of the backward-moving timeship register intervals of billions of years, we observe the great clusters of galaxies slowly approaching, then merging into one another. The galaxies, released from the embrace of their dissolving clusters, drift closer and closer together, while getting younger and younger.

Then, when the universe is a billion or so years old, the galaxies, one by one, here and there, swell up and become huge gaseous globes of hydrogen and helium. We stop and survey the scene with speculative eye.

Gradually the globes shrink and turn into bright lanterns. Their inner regions light up with swarms of first-born stars. Infalling gas descends between the stars, condensing and forming generations of new stars. In some globes the swirling gas settles and forms the rotating disks of spiral galaxies, where more stars precipitate. In other globes the gas continues to fall and settles finally into the nuclei of giant elliptical galaxies.

Said Hilaire Belloc in *More Beasts for Worse Children,*

> *Oh! let us never, never doubt*
> *What nobody is sure about!*

I forgot to remind you at the beginning that this journey occurs in the imagination, and consequently we see only what is known or thought to be known. Our timeship is really a dreamship. How galaxies form, and why they exist, we do not know, and whatever we venture to say on what nobody is sure about remains purely conjectural.

Of this only can we be sure: in a heroic age the sons of the morning beget the galaxies—the daughters of evening—whose starlit worlds create and nurture organic life.

□   □   □

"Now entertain conjecture of a time when creeping murmer and the poring dark fills the wide vessel of the universe," said the Bard. Across a "dark backward and abysm of time" we journey. Back through a turbulent darkness of the mesocosmic era to an age when the universe is tens of millions of years old. Little is known of this strange prenatal era of the galaxies. If more were known, we might understand how the expanding universe fragments into gaseous globes of galactic mass.

Perhaps flickering streaks of eerie light pierce the darkness, and gas writhes in the grip of tortuous magnetic fields; perhaps strewn everywhere are black holes of assorted masses forged in the paleocosmic eras of the early universe. Who knows?

Slowly the temperature rises, and the darkness melts into a glimmer of light. At an age of 1 million years the universe begins to glow red hot. "What dreadful hot weather we have! It keeps me in a continual state of inelegance," says a distinguished passenger, Jane Austen, quoting from one of her letters. Her remark reminds me that I have forgotten to put on the air conditioner.

We must now proceed with circumspection, taking shorter and shorter backward flights in time. At an age of 100,000 years the universe fills with yellow light almost as bright as the surface of the Sun. The density has risen to about a thousand atoms per thimble of volume, and though this may not seem much, it is more than a billion times the average density of the present universe and a thousand times more than the average density of our Galaxy.

At last we stand at the threshold of the early universe. From this epoch descends the cosmic radiation, cooled and enfeebled by expansion, that has been discovered in recent years.

□　□　□

In trepidation we cross the threshold and enter the so-called big bang, and find ourselves in a silent and serene world of incandescent light. Like ancient mariners we are the first that ever burst into this silent sea. We voyage across the comparatively long radiation era of the early universe, across an era that begins when the universe has an age of 1 second and ends when it has an age of 100,000 years. The dominant constituent in this era of cosmic youth is the intensely bright radiation flooding the universe.

We owe to George Gamow and his colleagues, Ralph Alpher and Robert Herman, the inspired idea of an early period dominated by radiation. Their theory of the radiation era, advanced in the late 1940s and early 1950s, was confirmed by the discovery in 1965 of the low-temperature cosmic radiation.

As we proceed back through the radiation era, the temperature rises, and the fiercely bright light around us soars in intensity. Glaring yellow light changes into dazzling white light that changes into searing ultraviolet light. When the temperature reaches 1 billion degrees, at a cosmic age of 3 minutes, and the light has transformed into an ocean of energetic X-rays, the universe becomes an immense nuclear reactor. A quarter of all matter (in the form of protons and neutrons) converts into helium, liberating a prodigious amount of energy. But the energy unlocked by the cosmic thermonuclear detonation falls a long way short of the energy already present, and the effect is only slight.

All the multitudes of stars, busily converting their hydrogen into helium over 10 billion years, have contributed only about one-tenth of the total amount of helium in the universe. The rest of the helium forms in the earliest stages of the radiation era. From the first 3 minutes comes also the deuterium that one day will be useful to us when scientists discover the trick of releasing energy on Earth by means of fusion.

The radiation era commences at about the time when the universe has an age of 1 second, a temperature of 10 billion degrees, and a total density of about 1 ton a thimbleful. We now leave the radiation era and enter the outlandish lepton era.

In this new era a deluge of lightweight particles fills the wide vessel of the universe. These newcomers, generated by the intense radiation, are the leptons, such as electrons, positrons, and their neutrinos. Soon other leptons—muons, antimuons, and their neutrinos—are generated and added to the dense lepton flood. From this era flee hosts of ghostly neutrinos that are with us to the present day. As numerous as the photons of the cosmic radiation, they roam freely everywhere, passing unimpeded through the Earth and other celestial bodies. These neutrinos, unlike the photons of the cosmic radiation, possibly possess a nonzero but small mass, and being so numerous, they may contribute most of the unseen mass of the present universe.

It is remarkable that with modern physics we can trace the history of the universe in broad outline back to the dawn of the lepton era, to a cosmic age of one-ten thousandth of a second, when the temperature is a few trillion degrees and the density near a billion tons a thimbleful.

Before us lies the extreme early universe, the unknown, mysterious proterocosmos. Our overworked air conditioner is now at full blast, and regrettably we can go no farther.

□ □ □

The dials in our timeship have registered shorter and shorter intervals, from millions of years to single years, then from seconds to fractions of milliseconds. Even though the steps taken have been progressively shorter, the cosmic scenery nonetheless has continually changed. The younger the universe, the faster it alters, and in the very early stages everything changes with extreme rapidity. In front of us the proterocosmos looms as a blur of action. Perhaps as much of cosmic history lies ahead as behind.

The complexity of the proterocosmos—the extreme early universe—reflects the complexity of the subatomic world, and our understanding of what happens, which is not very much, depends on the little we know of the world of subatomic particles.

Our timeship can go no farther. Standing at the dawn of the lepton era, sustained by little more than a spirit of speculative inquiry, we explore the proterocosmos by peering into a crystal ball.

□ □ □

Hadrons are the strongly interacting subatomic particles, such as protons, neutrons, and their antiparticles. Just before the onset of the lepton era, at a temperature of trillions of degrees and a density of billions of tons a thimbleful, the hadrons have their moment of glory as the dominant constituents of the universe.

Ordinary matter consists of normal particles (such as positive protons), and antimatter consists of antiparticles (such as negative antiprotons). The immense concentration of energy at this stage in the early universe creates particles and antiparticles as fast as they annihilate each other. Matter coexists with antimatter.

Arthur Schuster in 1898 predicted the existence of antimatter in a letter to the editor of the science journal *Nature* entitled "Potential Matter—A Holiday Dream." "When the year's work is over," he wrote, "and all sense of responsibility has left us, who has not occasionally set his fancy free to dream about the unknown, perhaps the unknowable." He went on to say:

> Surely something is wanting in our conception of the universe. We know positive and negative electricity, north and south magnetism, and why not some extra terrestrial matter related to terrestrial matter, as the source is to the sink. . . . Worlds may have formed of this stuff, with elements and compounds possessing identical properties with our own, indistinguishable in fact from them until they are brought into each other's vicinity.

He concluded with the words, "Astronomy, the oldest and most juvenile of the sciences, may still have some surprises in store. May anti-matter be commended to its care! . . . Do dreams ever come true?"

Paul Dirac in the late 1920s revived the possibility of antimatter in his work of uniting special relativity with the nascent theory of quantum mechanics. The positron (antiparticle of the electron) was discovered in 1932, and the antiproton in 1952; other antiparticles have since been discovered.

Schuster speculated that antimatter possesses antigravity. In this he erred. Particles and their antiparticles exhibit opposite aspects, such as positive and negative electric charge, but both kinds possess similar gravity and respond to gravity in similar ways.

Schuster also erred in supposing that worlds of antimatter are as common as worlds of matter. Matter and antimatter, when brought close together, annihilate each other with the release of energy and the emission of identifiable radiation. We see no convincing signs of this radiation heralding significant amounts of antimatter in the Galaxy, and none betraying the presence of maverick antigalaxies intermingling with galaxies in the extragalactic realm. The lost worlds of antimatter tell that our universe favors matter more than antimatter.

Among the hadron hordes of the early universe, before the onset of the lepton era, there exists a slight excess of matter over antimatter of about one part in ten billion. Hadrons and their antiparticles are constantly created and annihilated in the matrix of warring matter and antimatter. Annihilation overtakes creation as the temperature drops and the lepton era approaches. Matter and antimatter then disappear, and only the slight preexisting excess of matter survives. This slight excess—a fugitive remnant of the hadron hordes—is inherited by succeeding ages and constitutes all the matter of the present universe. The immense energy released by the annihilation of the hadrons goes eventually into the cosmic radiation.

We live in a topsy-turvy universe. The matter so vitally important in the composition of galaxies is the result of a small, freakish, and inconspicuous difference in the amounts of matter and antimatter in the proterocosmos. The present-day cosmic radiation, enfeebled by expansion and seemingly of little consequence, is the legacy of the fantastic energy of the proterocosmos. What once was inconspicuous has become important, and what once was important has become inconspicuous.

If matter and antimatter in the proterocosmos had equal abundance, they would have annihilated each other completely, and there would now be no galaxies, and we would not be here discussing the subject.

□　　□　　□

Hadrons are little worlds made cunningly from quarks. Because of the steady rise in density (we are still proceeding backward in time), the hadrons squeeze together more and more tightly and soon overlap. The hadronic boundaries dissolve, the quarks emerge, and the universe transforms into a dense sea of free quarks. We have crossed into the bizarre quark era that stretches away to almost the beginning of time.

For 50 years, following Maxwell's death in 1879, the universe was ruled by gravitational and electromagnetic forces. Sigmund Freud in horror exclaimed, "With these forces nature rises up against us, majestic, cruel, and inexorable." But on this subject Freud was out of his depth. Without gravitational forces there would be no galaxies, stars, and planets; without electromagnetic forces there would be no atoms, electricity, and light. Without either we could not exist.

In addition, we have nowadays the strong and weak forces of nature. Without the strong force, protons and neutrons would not combine into atomic nuclei heavier than hydrogen; without the weak force, protons would not combine to form deuterium, and hence hydrogen would not burn to helium, and the stars would not shine over long periods of time. Without either we could not exist. The four forces of nature rise up not against us but for us, making life possible in the universe.

Albert Einstein, after his success with general relativity, sought unsuccessfully to extend Maxwell's synthesis by combining gravity and electromagnetism within a unified geometric scheme. Sheldon Glashow, Abdus Salam, Steven Weinberg, and other physicists in more recent years, casting loose their moorings, have succeeded in showing that the electromagnetic and weak forces are dual aspects of a more fundamental electroweak force. At very high particle energies these dual aspects fuse and become indistinguishable. The electroweak force in unified form rules in the quark era. Nowadays it exhibits different guises—electromagnetic and weak—because particle energies are relatively low.

Proposed grand unified theories (referred to by the inelegant term GUTs) tie together the electroweak and strong forces into a hyperweak force. This force rules supreme before the onset of the quark era and disregards all distinctions between matter and antimatter. At ultrahigh energies the fundamental hyperweak force merrily transforms matter into antimatter and back again, blithely switching quarks into antiquarks, leptons into quarks, and vice versa, rolling the strong, electromagnetic, and weak forces into one. Utmost symmetry exists during the reign of the hyperweak force in the earliest moments of the proterocosmos.

The sharp distinction between matter and antimatter comes with the decline of the hyperweak force and the rise of the quarks and leptons at a time when the universe is only a trillion-trillion-trillionth of a second old. At the same time comes also the slight difference in the abundances of matter (quarks) and antimatter (antiquarks). Owing to the rapid expansion of the proterocosmos, the various interactions and their decay schemes get out of phase, and matter is whimsically favored by one-ten billionth more than antimatter.

The hyperweak force exhibits triple guises in the present universe. Maybe the hyperweak force still acts feebly in its original unified form. Because of this vestigial unity, matter perhaps cannot endure permanently and is slowly dissolving into radiation on a time scale of several million trillion trillion years. (Notice what happens on a time scale of a trillion trillion trillion seconds in the late universe matches what happens on a time scale of a trillion-trillion-trillionth of a second in the early universe.) The possible slow decay of matter—at a rate of one proton in each ton every year—is not entirely beyond the limits of experimental verification.

Experimentalists are rallying to detect whether our world is melting away, whether these "statelie courts, these sky-encountering walles," will eventually "evanish all like vapours in the aire," as predicted by the seventeenth-century poet William Stirling in the *Tragedie of Darius*.

Perhaps one day all four forces of nature will be happily combined into a supreme unity. Athwart the road to this desirable goal lies the perversity of gravity. Herculean attempts by the cognoscenti to bring general relativity into the fold of quantum mechanics have so far failed, and we have yet to see whether grandiose schemes of supersymmetry and supergravity can remove the roadblock.

□ □ □

Quantum black holes—the basic constituents of spacetime—are the ultimate and most energetic of all particles. They have a mass 10 billion billion times the mass of a proton (roughly that of a speck of dust), and a size 1 ten-billion-billionth the size of a proton. The foamlike texture of spacetime consists of frolicsome virtual quantum black holes, popping in and out of existence in unimaginable numbers, each savoring the joys of life for a Planck period. A Planck period equals 1 ten-million-trillion-trillion-trillionth of a second. These ultimate particles, weaving out the fabric of spacetime, have not been observed because of the extreme energy needed to lift them out of their virtual state. Heaven alone knows what would happen if we could create them; we might inadvertently conjure up the cosmogenic genie, tear apart the fabric of spacetime, and burst open the universe with the eruption of a new cosmos.

The likely existence of quantum black holes indicates that we can peer into our crystal ball back to a time when the universe has an age of one Planck period and a density of 10 followed by 93 zeros times the density of ordinary solid matter. The entire observable universe of the present day fits into a size smaller than a hydrogen atom. If the universe is closed, this size would equal roughly the circumnavigation distance; but if the universe is open and of infinite extent, then it is also of infinite extent in the beginning, for infinity lacks limits and cannot be compressed.

Our journey back through the proterocosmos comes to a final halt at an impenetrable barrier where the universe has an age equal to a Planck period. Around us lies a foam of inconceivable chaos in which time and space are torn into discontinuities of cosmic magnitude. We can go no farther. An orderly historical sequence of events has ceased to exist, and past and future have become meaningless. Here, in the realm of quantum cosmology—the chaosmos—lie the secrets that foretell the architecture of the universe.

□ □ □

We see the universe stretching away much the same in all directions. But our location in space is not particularly privileged, and probably the universe stretches away much the same as seen from any other place.

Probably, all places are alike, and the universe is therefore homogeneous. A rational universe requires that everything has a natural explanation. How, then, can we explain homogeneity?

In a static and eternal universe ample opportunity exists for all places to interact and adjust into uniformity. But widely separated places in an expanding universe generally lack opportunity because of insufficient time.

The Hubble sphere swells as the universe expands, and we see more and more the remote parts of the cosmos that previously were unknown. The remote parts, when revealed, turn out to be the same as the known parts. We are inclined to think that the cause of this homogeneity, like all other causes, can be found in the past. But when we journey backward in time, seeking the origin of homogeneity, the Hubble sphere shrinks, and we see less and less of the cosmos. When the universe is a million years old, we see a distance of roughly only a million light-years, and when a second old, roughly only a light-second. The range of communication and interaction diminishes, and at the Planck epoch the range extends only to neighboring particles. In the past the remote parts of the universe that we see nowadays did not know that one another existed. How, then, could they adjust into conformity?

One way of explaining homogeneity, recently proposed by Alan Guth at the Massachusetts Institute of Technology, supposes that the extreme early universe inflates with a Hubble sphere of constant size. The observed universe of today originates inside this Hubble sphere, thus affording the ample opportunity necessary for interaction and adjustment into uniformity. This age of inflation exists before the onset of the quark era.

The inflationary era exists because of a weird cosmic tension (the opposite of pressure) that comes into play as a possible consequence of the grand unified theory. With the decline of the hyperweak force at the end of the inflationary era, the cosmic tension dissipates into a dense and hot sea of quarks and leptons.

In one of his philosophical moods Arthur Eddington wrote, "the notion of a beginning of the present order of Nature is repugnant to me." Other scientists have shared Eddington's dislike of a cosmic beginning. What they feared—a revival of the melodrama of Babylonian mythology—has to some extent come true. The awesome "Big Bang" has replaced the dreaded Marduk, conqueror of the monster of primeval chaos, Tiamat.

Whatever one's philosophical view, the stubborn fact remains that the afterglow of an inchoate cosmos, dense and hot, has been discovered. The cosmos of long, long ago, known to the scholarly and discriminating as the early universe, is the popularly acclaimed "Big Bang."

Fred Hoyle coined the vogue name big bang in his sensational BBC lectures *The Nature of the Universe* of 1950. This catchy locution, with its mythic overtones, has misled many persons into the false belief that the universe originates as an explosion at a point in space. Hoyle to some degree can be forgiven; in his missionary zeal on behalf of the steady-state universe, he ridiculed the rival theory of an evolving universe by using the pejorative term big bang. But it backfired. Less forgivable, though understandable, is its enthusiastic adoption by marketplace purveyors of pop cosmology.

From our journey into the distant past we have learned an important lesson. To reach the early universe we remain where we are in space and travel back in time.

At any moment the universe appears much the same everywhere in space. We cannot go to some other place in space and find the early universe sitting there. We reach it by traveling not in a spaceship but in a timeship. We travel not synchronically but diachronically. This vital truth must be stressed to counter the mischievous connotations of the name big bang. The words big bang erroneously imply that the universe begins as an explosion at a point in space.

A deceived person might ask: Surely not everything need be inside the big bang? Surely we can stay outside at a safe distance and watch it explode? But the universe does not explode at a point in space. The so-called big bang fills all space, actually is the early universe, and in no circumstances can we get outside. The universe contains all space and is not contained within space.

"It's a poor sort of memory that only works backward," opined the White Queen. While we still have the timeship, we might as well try to explore the future. For some unknown reason the machine refuses to stop anywhere in the immediate future.

We leap forward in time. Again the dials record the passage of eons. After 5 billion years the Sun swells into a red giant and soon fades into a white dwarf. Terrestrial life has perished or fled elsewhere. After tens of billions of years guttering stars faintly illumine the fall of cosmic night.

Before us lie alternate scenarios: the universe either collapses and ends with the return of primeval chaos or expands endlessly in vacuous darkness: either cremation on a flaming pyre or burial in a somber vault.

Consider the first alternative in which the universe ends in a blaze of light. After 40 or so billion years expansion ceases and collapse commences. The galaxies begin to approach one another, and after a further 40 or so billion years, they arrive back where they are at present. Roughly 10 billion years remain till the end of time.

The great dissolution begins. First, the clusters merge together, then the galaxies themselves overlap and dissolve. The universe now consists mainly of old stars (dwarfs and neutron stars) immersed in a commotion of gas. Once more we proceed with circumspection, taking shorter and shorter flights in time, plunging into the antibang.

The big squeeze starts in earnest, and the ensuing uproar defies description. Edgar Allan Poe more than a hundred years ago portrayed in *Eureka* the end of a collapsing universe. "Then, amid unfathomable abysses, will be glaring unimaginable suns. But all this will be merely a climactic magnificence foreboding the great End." All astronomical systems are dismantled, and all remnants of life obliterated. The stars accelerate and career about helter-skelter at higher and higher speeds, approaching the speed of light; a few collide headlong and erupt, but most wear away to nothing, leaving sparkling trails as they zip through the tumult.

After a total lifespan of about 100 billion years the former brilliance returns and the universe reverts to primordial chaos. What happens then we do not know; perhaps a similar or an entirely different universe rises from the ashes of the old.

Consider the second alternative, in which the universe expands forever and ends not with a bang but a whimper. Galaxies illumined by occasional flickers of light voyage farther and farther apart in the dark abysm of time. The dials in our timeship whirl faster, and after trillions of years all is dead in a funereal world of darkness.

Yet powerful and unremitting agencies are at work. Dark stars flee and take refuge on the outskirts of shrinking galaxies; dark galaxies flee and take refuge in the widening abysses between shrinking clusters. A steady outpouring of gravitational waves augments the slow collapse of stellar and galactic systems.

Hitherto we have merely dawdled on our journey. I set the timeship into overdrive, and we leap forward across a million trillion trillion years, and find that most matter is huddled into large black holes. On this vast time scale the hyperweak force exacts its toll. Stars and galaxies that have escaped capture by black holes melt away into radiation, and "evanish all like vapours in the aire."

In the black universe only black holes exist. Across unspeakable spans of time we now see the black holes themselves melting away.

Stephen Hawking of Cambridge University has shown that black holes emit radiation quantum mechanically. In the same wavelike manner that protons penetrate each other's electrical repulsion barriers, particles and radiation tunnel through the surfaces of black holes and escape into the outside world. This nonclassical defiance of the classical world occurs most easily at wavelengths the size of black holes. Because of this emission of energy, consisting mainly of photons and neutrinos, the black holes continually lose mass.

A black hole of solar mass evaporates away in a time, measured in years, of 10 followed by 65 zeros. Black holes of galactic mass last much longer, and their lifespans are measured in googols of years. A googol—a term made famous by Edward Kasner and invented by his 9-year-old nephew—stands for 10 followed by 100 zeros. When our time machine is recording intervals of googols of years, the largest black holes have vanished. The black universe has become the blank universe. What remains is little more than feeble radiation forever becoming feebler.

Faster and faster we travel on a journey without end in an insane accelerating timeship that has jammed and refuses to stop. Its dials measure the passage of googolplexes of years, where a googolplex is 10 followed by a googol of zeros. Googoogolplexes (my juvenile expression for 10 followed by a googolplex of zeros) of googoogolplexes of years pass in pitch-black vacuity, and still eternity lies ahead.

☐ ☐ ☐

We do not know whether the universe will recollapse or endlessly expand. The simplest cosmological models show that a closed universe expands and collapses, and an open universe endlessly expands. The observations by astronomers are still too imprecise to determine which possibility is more likely. The results obtained have been repeatedly revised, and only the brave dare rely on them with confidence. The observations are interpreted usually with the simplest models, because more elaborate versions are at present beyond hope of verification.

An eternal future offends against one's belief that the cosmos has rational unity. In a philosophical mood I like to think that a universe having a semblance of unity must contain finite time as well as finite space. I share Emily Dickinson's modest prayer:

> And so upon this wise I prayed—
> Great Spirit give to me
> A heaven not so large as yours
> But large enough for me.

Only a cosmic jester could perpetrate eternity and infinity.

Try to imagine if you can an infinite expanse of space. Out there, googols of light-years away, the cosmic scenery is just the same as here.

Even if it is not the same, and the universe is inhomogeneous, every possible variation over googolplexes and googoogolplexes of light-years is monotonously repeated. On sufficiently vast scales an infinite universe is always homogeneous. Endless repetition of identical things occurs in an endless universe.

Our Solar System consists of a large but finite number of atoms. These atoms may be arranged into an enormous though finite number of different configurations. Each distinguishable configuration of this multiworld ensemble has finite probability.

Given an infinite number of solar systems, as there must be in an infinitely large universe, every configuration of finite probability is repeated an infinite number of times. Nothing is unique. Out there exists an infinite number of identical solar systems, with identical earths, with identical human populations. Each of us at this moment is doing the same thing in an infinite number of places. What can be the point of all this futility when once is often more than enough? If eternity is silliness, then infinity of space is sheer madness.

We have a global view of the physical universe. A fascinating view that may be shaken into kaleidoscopic patterns. We marvel at these many patterns, setting aside most as implausible, selecting a few, usually the simplest, consistent with what we know. We believe the universe to be at least as complex as anything it contains. Yet a single subatomic particle is incredibly complex; can the universe that supposedly makes sense of it be any simpler?

Do dreams ever come true? Does our universe really contain us— the dreamers?

Part 3

The
Cloud
of
Unknowing

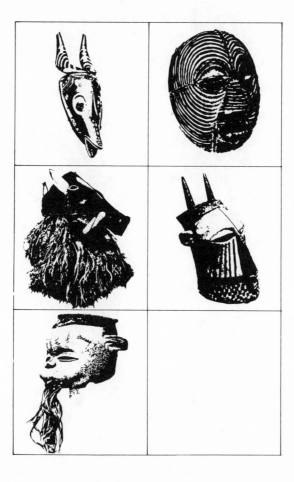

13

The Witch Universe

Francis Bacon, English courtier and statesman of the late sixteenth and early seventeenth centuries, promoted a philosophy of empirical science and declared, "let every student of nature take this as his rule that whatever the mind seizes upon with particular satisfaction is to be held in suspicion." His belief in empirical methods of investigation assured him that witches existed.

Let us look at the witch universe that this incredulous and illustrious man lived in.

☐ ☐ ☐

Tracing the development of ideas in the long Middle Ages leads the historian into a labyrinth of bewildering beliefs. The works of Jabir ibn Haiyan, court physician in the eighth century to Harun al-Rashid (caliph of Baghdad famed in *The Thousand and One Nights*), became widely known for their medical lore and learned alchemy. Jabir was later latinized into Geber, and because of the rigmarole and obfuscation of the numerous works attributed to him, the word Geberish eventually became gibberish.

In the Middle Ages the telluric elements of earth, water, air, and fire exhibited the qualities of cold, wet, dry, and hot. By erudite arguments the elements accounted for bodily humors of melancholy, phlegm, choler, and blood, and in marvelous manner corresponded with the characteristics of creation, fall, redemption, and judgment. Acts of the will were governed by God, acts of the intellect by angels, and acts of the body by celestial orbs. Each person possessed a daemon or genius who acted as a guiding spirit. Less remarkably, the ten wits comprised the five outer senses of sight, hearing, smell, taste, and touch, and the five inner senses of memory, thought, imagination, instinct, and common sense.

Metals according to alchemists and astrologers possessed intimate relations with the celestial bodies: silver with the Moon, quicksilver with Mercury, copper with Venus, gold with the Sun, iron with Mars, tin with Jupiter, and lead with Saturn. The laboratorium of the medieval alchemist was lavishly equipped with vessels, vials, urinals, alembics, descensories, sublimatories, and various paraphernalia, with furnaces and other contraptions installed for research in calcination, sublimation, distillation, condensation, and solification. Chaucer in *The Canon's Yeoman's Tale* gives us a peep into the medieval laboratory, and we see how richly it is stocked with alkali, arsenic, brimstone, sal ammoniac, salpeter, quicksilver, vitriol, and herbs of every kind. The abracadabra, incantations, obscene prescriptions, and the rest seem to us pure gibberish.

The consuming passion of the alchemists, not unlike our modern quest for the cure of cancer, was the search for the philosopher's stone, which when discovered would transmute base metals into gold, restore lost youth, and grant eternal life on Earth.

□   □   □

Kings believed that by augmenting their power they implemented the divine intention. As their authority grew in the late Middle Ages the power of the feudal nobility waned, and European nations drifted apart, divided by conflicting national interests.

Wider horizons, strange lands, new lifestyles, and novel schemes of thought demanded a more capacious world view, and the medieval universe was soon bursting at the seams. Its disruption in the late Middle Ages and disintegration in the Renaissance was accompanied in Western Europe by widespread unrest and social upheaval. Warfare punctuated by plagues erupted and became a way of life, arresting the population growth.

The Church ceased to have a civilizing influence and seemed bent on wrecking all it had accomplished. Assertions that Jesus was a poor man affronted rich prelates and were condemned as heresy by popes. Corrupted by power, riddled with simony, demoralized by crusades, and lost in the casuistry that glorious ends justified inglorious means, the Church was incapable of implementing long-overdue reforms. In vain Erasmus and enlightened humanists endeavored to moderate social tensions with counsels of forebearance. The Reformation and Counter Reformation armies marched and countermarched amidst the ruins of the medieval universe.

Arts, classics, drama, and poetry became havens in which distraught sections of the public sought shelter from a grim and stormy reality. Painters in brilliant colors glorified the human figure against mythical backgrounds; poets in wondrous words rhapsodized on themes of passionate love; dramatists in imaginative fantasies catered to enthralled audiences. Thousands of plays opened up escapist worlds, and one-tenth of the adult population in London could be found of an afternoon at the theaters.

It was an age of disillusion, of make-believe, of paradise lost in a ravaged world. We call it the Renaissance and forget that the engines of art are fueled by the distillates of anguish.

Not unexpectedly, out of the despair of a world in disarray came a pathological desire to find scapegoats. First the Jews, who denied the godhead of Christ, had poisoned the wells and caused the plagues. Then the witches, with their black arts, had formed a conspiracy with the Devil and caused all the misfortunes of the sixteenth and seventeenth centuries. The Church, backed by secular authority, led the campaign of persecution.

The crusades had shown that wars waged against external enemies were an effective way of propping up absolute rule by distracting the public. After the collapse of the Christian venture in the Holy Land, the grip of central authority was maintained and tightened by stirring up fears of internal enemies.

Witchcraft until the thirteenth century consisted mostly of superstitions and folk beliefs that added variety to the lives of both highborn and lowborn, and much the same as today accounted for the fortunes of life when other explanations seemed unconvincing. The arts and crafts of witchery covered a broad spectrum, ranging from the benign to the malign; from healing, dispensing herbal preparations, midwifery, charms, spells, love potions, to casting the evil eye, mutilating waxen images, invoking the dead, contriving miscarriages and impotency, conjuring up storms and floods, causing diseases, even death. To cultured clerics, witchcraft was deplorable, and much of it seemed ludicrous and unrelated to Christianity.

The witch hunt began in the late Middle Ages and developed into the witch craze of the Renaissance. By the sixteenth century witchcraft had become a reality of supreme significance, and all lingering doubts about its diabolical nature had vanished. Even folk dances and festivals of pre-Christian origin were associated with the machinations of witch-craft and devilry. Biblical texts on the subject of witches and demons (with the express command "thou shalt not suffer a witch to live") sup-plied the principles of a new universe.

The witch universe was a dark, inverted image of Christianity. As seen in this demented universe, a hideous conspiracy with the Devil had mantled the Earth in a sinister twilight. The Lords of Light and Darkness were at last face to face, locked in a life-and-death struggle. In the desperate war of the worlds that ensued, involving all sections of the public and mobilizing all resources of church and state, fires were stoked, torture chambers equipped with the latest weapons, and monkish armies recruited to fight against an invasion of horrifying demons. The blasphemous, sacrilegious, cannabalistic witches were devil-worshiping demons; they flew through the air on besoms or grisly beasts to clan-destine covens, assumed grotesque animal shapes, killed and ate young children, desecrated the Cross, and were associated in the minds of cel-ibate priests with the most obscene sexual practices. Such was the ca-lamitous state of European society after the collapse of the medieval uni-verse.

□   □   □

In the years 1258 and 1260 Pope Alexander IV issued two bulls bid-ding first the Franciscans and then the Dominicans to refrain from judg-ing witchcraft unless heresy was amply demonstrated. Two centuries later the outlook had totally changed: the arts of witchcraft and the most heinous forms of heresy were one and the same.

The fables of countryfolk and townfolk were little more than vestiges of old pagan beliefs about hobgoblins, fairies, and the like. Zealous clerics in the late Middle Ages, increasingly fearful of nonconformity in a society of growing complexity, exaggerated these popular credulities and with fevered imagination invested them with elaborate sophistication. They summoned up the demonological panoply of the ancient world and projected it into the minds of the populace.

Whenever the inquisitors questioned the members of a dissenting sect or dissident cult—such as the Cathars and the Waldenses—they found that all dissenters and dissidents had similar revolting vices: they killed babies by tossing them from one person to another, then ate their bodies, had identical vicious and horrid habits, held sexual orgies at nocturnal meetings, and called up the Devil, who appeared as either a young man or a noisome beast. Simple and uneducated persons were found guilty of abominable practices that scholars trawled from mythology and dredged up from the history of olden times, of which the population previously had not the slightest knowledge. Peasants, distinguished persons, and even the proud Knights Templars were the innocent victims of a demonizing mania that gathered momentum and swept through Europe.

The method of interrogation, legitimized by the popes, consisted of questioning the accused victim about his heretical and gruesome habits, which were described to him in lurid detail. When the accused denied such behavior, torture succeeded sooner or later in extracting the required admission. Torture could be mitigated by confessing the names of other heretics. In this way the demon fantasies rife in the overcharged minds of the inquisitors became established truths, and society accepted the reality of a paranoid witch universe strenuously foisted on it by self-deluded intellectuals.

◻ ◻ ◻

The witch craze began with the bull of Pope Innocent VIII issued in 1484 that deplored the widespread depravities of witches. In bitter sorrow the pope enumerated the appalling enormities of witchcraft:

> Many persons, of both sexes, unmindful of their own salvation and straying from the Catholic Faith, have abandoned themselves to devils, incubi and succubi, and by their incantations, spells, conjurations, and other accursed charms and crafts, enormities and horrid offences, have slain infants yet in the mother's womb . . . these wretches furthermore afflict and torment men and women, beasts of burthen, herd-beasts, as well as animals of other kinds, with terrible and piteous pains and sore diseases, both internal and external; they hinder men from performing the sexual act and women from conceiving . . . and at the instigation of the Enemy of Mankind they do not shrink from committing and perpetrating the foulest abominations and filthiest excesses to the deadly peril of their own souls, whereby they outrage the Divine Majesty and are a cause of scandal and danger to very many.

In this bull the pope appointed "our dear sons Heinrich Kramer and James Sprenger, professors of theology, . . . as Inquisitors of these heretical pravities," to go forth into northern Germany and wherever else they could find "the disease of heresy and other turpitudes diffusing their poison to the destruction of many innocent souls." The pope empowered them to punish and condemn the offenders.

Kramer and Sprenger were distinguished on account of their vigilant defense of Christian society, inquisitorial zeal, and diligent burning of heretical witches. They sallied forth at the pope's command and 2 to 3 years later reported the results of their investigations. Their report was published under the title *Malleus Maleficarum* or *The Hammer of Witches* (the full title reads, *The Hammer of Witches, which Destroyeth Witches and their Heresy as with a Two-edged Sword*). Some details of this report deserve quoting, for in restrained and moderate language (by the standards of those times) the cosmologists Kramer and Sprenger reveal to us a picture of the witch universe in the making.

They find, as reported in the *Malleus Maleficarum*, that the towns and countryside swarm with astrologers and sorcerers consulted by all classes, low and high, and that political plottings and royal schemings are everywhere associated with the black arts; they notice "in this twilight and evening of the world, when sin is flourishing on every side, and in every place, when charity is growing cold, and the evil of witches and their iniquities superabound," how evident it has become that "witches and the Devil always work together, and that insofar as these matters are concerned, the one can do nothing without the aid and assistance of the other"; and for the benefit of the reader they outline the various branches of witchcraft, present numerous anecdotal accounts of witchcraft, and specify in detail the infamies and evils practiced by witches; thus of the forty-one witches burned in 1485 at Como, they affirm that these creatures had intercourse with the Devil, and "this matter is fully substantiated by eye-witnesses, by hearsay, and the testimony of credible witnesses"; for the witches have lewd practices in order that they may increase to the detriment of the Faith, and in their homes are visited by the Devil who copulates with the he-witches as a succubus and with the she-witches as an incubus, and the semen taken from one and given to the other breeds more and even worse witches; women in particular are vulnerable to the wiles of the Devil, for they are feeble in mind, credulous, more carnal than men, and liars by nature, and the works of the blessed saints and holy fathers attest that "all witchcraft comes from carnal lust, which is in women insatiable" (for what "else is woman but a foe to friendship, an inescapable punishment, a necessary evil, a natural temptation, a desirable calamity, a domestic danger, a delectable detriment, an evil of nature, painted with fair colours!"); and note that one of the aims of witches is to kill as many children as possible before Baptism, debarring them from Heaven and thus preventing the Elect from reaching the final number, thereby delaying the Day of Judgment in this twilight age; witches eat children, and from the bodies of children—often stolen from graves—they concoct obscene unguents essential in their transportation and transformations; witches stir up hailstorms, raise tempests, and cause thunderbolts to blast men and beasts, and are an abomination

Three witches burning, Germany 1555.

and the worst of heretics; those bishops and all rulers who fail to essay their utmost in stamping out witchcraft must be judged as abettors and punished as heretics; and "any witness may come forward" and the "accused person, whatever his rank or position, upon such an accusation may be put to the torture, and he who is found guilty, let him be racked, let him suffer all other tortures prescribed by law in order that he be punished in proportion to his offences"; and nowadays instead of being thrown to wild beasts as in olden days, "they are burnt at the stake and probably this is because the majority of them are women"; if the accused demands to confront her accusers, or have a legal advocate, it is up to the judge to decide, and he must take into account that such matters always delay and confuse the proceedings, and must weigh the possibility that the accused speaks with the voice of the Devil; each witch must be stripped ("by honest women of good reputation"), shaven of all hair, and put to the question while naked; at first, in justice, she must be questioned lightly without shedding blood, even though moderation in questioning is always fallacious and generally ineffective; she must be questioned about her intercourse with the Devil, the infants killed and eaten, and other matters of which she is justly accused; note that witches frequently commit suicide between the periods of examination and must be prevented by shackling, for they are induced to do so by the Devil; when witches confess they are prompted always by a divine impulse from an Angel, but when they plead innocence or do not confess or have "the evil gift of silence that is the constant bane of judges" they are so impelled by the Devil, and the interrogation must continue; witches who cry out under investigation give voice to devils, yet those who cannot weep are manifestly also in the possession of devils; on the other hand, weeping may itself be an artifice of the Devil seeking to avert justice by creating pity; those who expire during the examination are saved by the Devil for the evil purpose of escaping full confession; an inquisitor or judge may facilitate his proceedings by promising the accused her life if she fully confess and name her evil heretical collaborators, though under no circumstances must such promises be kept, and may be made with "the mental reservation he will be merciful to himself or the state," for "whatever is done for the safety of the state is merciful"; when a confession has been extracted, the witch must then be handed over to the secular authority for execution, because the Church cannot exact the ultimate penalty; divine providence has arranged that all inquisitors and judges are immune to the powers of witches, and they should therefore never be deterred in their proceedings; and inquisitors and judges must understand that witches are the incarnations of Evil who have made a compact with the Devil and are an affront to the Terrible Judge, and must remember that witches, "however much they are penitent and return to the faith, must not be punished like other heretics with lifelong imprisonment, but must suffer the extreme penalty," and inquisitors and judges must bear in mind that if those "who counterfeit money are summarily put to death, how much more must they who counterfeit the Faith."

The *Malleus Maleficarum* possesses a hypnotic power. The aghast and sickened reader, after returning from the nightmare universe of Kramer and Sprenger, will never again be quite the same person. Only those who claim that their universe is the Universe, because it has been verified, should read this monstrous work of verification.

The point at issue is not whether the things reported were actually happening—which is for the sane reader to judge—but whether they were believed to be happening, and on this point there is not a scintilla of doubt. *Malleus Maleficarum* in the fifteenth century revealed the stark contours of reality. While numerous supplementary texts described, explained, and filled in the details, the *Malleus* became the standard textbook, the authoritative *Principia* of a new universe. For more than two centuries *The Witch Hammer* had enormous influence and was used by Catholics and Protestants alike. In constant use, in the hands of every scholar and on the desk of every official, it dispelled all reservations concerning the reality of witches; it spurred the fainthearted and urged to even greater efforts the zealous in their constant battle against witchcraft.

A good Christian had no recourse other than to support openly the Stalinistic eradication of the vile forces subverting society, in fact, had no choice other than to bear witness whenever necessary against the insidious heretical leanings and therefore witchcraft tendencies in spouse, parent, brother, sister, son, daughter, or other relative, and of course, neighbor.

□   □   □

In the Renaissance, with its supposed "rebirth in the nobility of the human spirit," every scholar and peasant shared the same beliefs concerning stereotype witches, was aware of the same fixed elements in the composition of witchcraft, and knew that witches flew through the air to their sabbats held in caves, on mountaintops, and other eerie places, where they gathered with an insatiable lust for young flesh at cannabalistic feasts, cavorted to macabre music, indulged in sexual orgies, parodied Christian ritual, renounced God, and worshiped the Evil One who appeared before them in dreadful guises.

And the awful fact was that whenever you found one witch and used the just and proper instruments of inquiry, you inevitably found many others. Their numbers multiplied and seemed without limit. Male and female witches, and their evilly spawned children, were consumed by fire in mounting numbers, and still they multiplied. Trevor-Roper in *The European Witch Craze* depicts the scene:

> All Christendom, it seems, is at the mercy of these horrifying creatures. Countries in which they had previously been unknown are now suddenly found to be swarming with them, and the closer we look, the more of them we find. All contemporary observers agree that they are multiplying at an incredible rate. They have acquired powers hitherto unknown, a complex international organization and social habits of indecent sophistication. Some of the most powerful minds of the time turn from human sciences to explore this newly discovered continent, this America of the spiritual world.

The details they discovered are constantly and amply confirmed by other research workers—experimentalists in confessional and torture chamber, theorists in library and cloister—leaving the facts still more securely established and the prospect even more alarming than before. Instead of being stamped out, the witches increased at a frightening rate, until the whole of Christendom seemed about to be overwhelmed by the marshaled forces of triumphant evil.

To protest in any way against witch hunting as inhuman in a time of dire emergency was sheer lunacy, condemned by the popes as bewitchment and the result of consorting with devils.

Who were the witches? They were the old and lonely, mostly female, the ugly and crippled, the weak and sick in mind, the insane, the objects of envy, the hated who could be spited by false testimony, the accused in village feuds, rootless peasants, strangers, dissenters, as well as charlatans, poisoners, and other malefactors. Their protests of innocence had no effect other than to confirm their guilt; once accused, confession and the naming of confederates was the only release from torture, followed by incineration.

It was an age that believed in witchcraft, and the accused shared the prevailing beliefs. Many of the accused suffered from hallucinations and thought they possessed the alleged witchcraft powers. Incarcerated under dehumanizing conditions, fed with an enfeebling diet, then put to the question, many if not most were induced to think they were guilty of the alleged offenses. The number of victims burned at the stake is unknown; the total is variously estimated from hundreds of thousands to millions, and was probably between half a million and one million.

Not the humanities, not religion, but the sciences saved Europe from the mad witch universe of the Renaissance. While contributing to the demolition of the medieval universe, the sciences were at the same time reaching out to a vastly different kind of universe, a universe more capable of defining the limits of human control over nature, in which the gibberish of demonology would become incredible and seem as if it could never have happened.

The emergence of science, says Herbert Butterfield in *The Origins of Science*, "outshines everything since the rise of Christianity and reduces the Renaissance and Reformation to the rank of mere episodes," and "looms so large as the real origin both of the modern world and the modern mentality that our customary periodisation of European history has become an anachronism and an encumbrance." Science in the Middle Ages with its handmaiden of technology established the framework and social structure of the medieval universe; subsequent developments in science overthrew the medieval universe and reached out to the more enlightened universe of the Age of Reason.

The Renaissance, like the Dark Ages, is a historical fiction invented in the eighteenth century. The supposed renaissance was a disordered interlude between sane universes, a bedlam of distraught world pictures terrorized by a witch universe created by leaders with fear-crazed minds, an age in thralldom to a mad universe on the rampage, which would have destroyed European society but for the intervention of science.

The general idea that we live in a known Universe, whose facts are moderately secure and much the same for everybody, tempts us to believe that witches and a witch conspiracy actually existed, thus justifying to some degree the witch craze. The people in the Renaissance, including witch persecutors like Martin Luther and Francis Bacon, were no more foolish than we; and if the contemporary facts lent credence to their belief in witches, surely the same facts still hold water.

But the known world in which we live is not the Universe, only a universe of our own contriving. The contemporary facts support an immense edifice of thought that in turn selects and interrelates those facts. When the facts are reinterpreted, the edifice crashes down, and the so-called facts then take on new meaning, and may even be discarded and replaced with new facts or old ones seen in fresh light.

The purpose of this chapter is to try to clarify two points of cosmological interest. The first is that we should judge the people of other societies in the context of their universes and not as if they lived in our universe. We must imagine the facts not as seen by us but as seen by them, and we must realize that our facts and their facts belong to distinctly different universes.

The mild-mannered student after years of indoctrination takes charge of the gas chambers in a Nazi concentration camp and a decade later turns up as a jovial shopkeeper. The student, exterminator, and shopkeeper are not the same man living in a known invariant Universe, but different men living in different universes. A person in an extraordinary universe is an extraordinary person by our standards and does not behave and think in ways we regard as ordinary. You and I in the mad witch universe might easily be zealous witch hunters engaged in an unremitting campaign against witchcraft, and yet still regard ourselves as normal and ordinary persons. With equal ease we might believe ourselves to be witches. We would conform to the climate of opinion in that universe.

The second point is that the universe in which we live is a real and actual world that rationalizes much of our experience and embraces most of current knowledge. We see the past, the present, and the future in the context of that universe. We believe in our universe, and the supporting facts are clear for us all to see. Whether the facts are liked or not does not really matter; they are there; that is the way they are; they cannot be altered. The universe in which a person lives is always real—verifiably real—and is not a hallucination, or if it is, then it is a collective hallucination of social magnitude that itself judges what is sane and insane, what is real and unreal. This second point is of great importance, and in the next few pages I discuss it at greater length.

☐ ☐ ☐

No one knows what constitutes the scientific method. Francis Bacon, a condemner and persecutor of witches who contributed little to science other than his weighty opinions, declared that science is wholly empirical and fully inductive.

Empiricism is the belief that the scientific method consists of attentive minds recording observations and drawing logically obvious conclusions. It is a truth-seeking method of inquiry that had for Bacon the incidental merit of confirming his belief in the reality of witchcraft. Seeing is believing, which Bacon emphasized, but also believing is seeing, which he failed to realize. Science demands the exercise of disciplined imagination before, during, and after an observation and is much more than mere empiricism.

The philosopher Karl Popper, with persuasive skills rivaling those of Bacon in his *Novum Organum*, has in recent years argued that science consists of whatever may be falsified by observation. Scientific facts have the distinction that they can be checked by observation and tested by experimentation. They are vulnerable to falsification—hence are falsifiable—and must stand on their own feet, unsupported by faith, hope, and charity. Popper is widely esteemed for his work on the scientific method and is acclaimed as the greatest philosopher of science that has ever been, and there is no more to science than its method, and there is no more to its method than Popper has said.

According to Popper, the difference between science and nonscience is that science is falsifiable, whereas nonscience is unfalsifiable. Facts and statements are unscientific if they are invulnerable to falsification and cannot be verified by repeated testing. The statement "angels exist" is unscientific, because in the modern universe it is unfalsifiable. You may believe in angels if you wish, but you cannot claim their existence is a scientific fact. All unfalsifiable elements of knowledge are either mythical, fanciful, or metaphysical.

Thus "all swans are white" is a scientific statement, acceptable until reports reach us of the existence of black swans in Australia. But "all angels have wings" is not a scientific statement, because however long we wait in the modern physical universe, there can be no reliable reports of observations to the contrary.

Popper's remarkable demarcation of science is now accepted by many scientists and philosophers. When taken literally, however, his criterion of what constitutes science condemns astronomy, biological evolution, and cosmology as subjects of little scientific merit; bluntly stated, they are unscientific and on the level of present-day astrology, creationism, and oriental mysticism. Apparently, when Popper proposed the theory of falsification, he failed to foresee all the consequences of his argument. Biological evolution, astronomy, and cosmology, which deal, respectively, with the antiquity of life, the longevity of stars, and the great age of the universe, have come under renewed attack from biblical fundamentalists now armed with Popper's definition of science, who boldly claim that if these subjects of unfalsifiable knowledge are accepted as science, then biblical creationism is science also, deserving equal attention in our schools.

"Falsifiable facts" are those facts that if wrong can be proved to be wrong. The trouble with the claim that science consists only of falsifiable facts is that these facts dance different ways when the theoretical tunes change and are rarely as simple and concrete as the stone attacked by Dr. Johnson. A science such as astronomy consists of interlaced elements of knowledge that are not always independently falsifiable; it is a self-adjusting system of advancing knowledge, governed by reason, not by faith as in astrology.

In 1932 at a meeting of the International Astronomical Union held at Cambridge, Massachusetts, the great astronomer and cosmologist Arthur Eddington declared, *"There are no purely observational facts about the heavenly bodies"* (quoted in *The Expanding Universe* with his italics). The observations by astronomers on Earth and their conclusions drawn from these observations acquire particular meaning through a long chain of argumentation in which the links are forged by physics and mathematics. What is observed here in space leads to an understanding of the structure and evolution of stars thousands and millions of light-years away.

Similarly with biological evolution: the links in a long chain of argumentation are forged by biological and physical sciences, and what is observed now in time leads to an understanding of the structure and evolution of organisms thousands and millions of millennia in the past. The claim by creationists that biological evolution is unfalsifiable and therefore not a science because we cannot travel back in time is as inept as saying that astronomy is unfalsifiable and therefore not a science because we cannot travel out in space.

The new Baconian philosophy restricts us to our backyards. The rest of knowledge dealing with things beyond our immediate experience is unfalsifiable and therefore metaphysical. Scientists who accept Popper's demarcation theory of science have not asked themselves whether the theory is itself falsifiable.

All universes are falsifiable within their own terms of reference. The witch universe, which you must agree is an unlikely example, serves as an illustration. If I can show that it was falsifiable within its own terms of reference, and if you agree that it was not scientific by modern standards, then Popper's theory of what demarcates science is falsified.

The witch universe was vividly real in the minds of those who lived in the Renaissance. It was believed to be true by inquisitors, jurists, victims, and every section of the public. It had its graduate textbook, *Malleus Maleficarum,* its phenomena were thoroughly investigated, and its facts repeatedly verified with the instruments of the torture chamber. To say that these instruments of inquiry are not those used in a modern laboratory is quite irrelevant; at that time they were deemed to be appropriate and truth seeking. (If you insist on this point, I must ask you what will be the instruments of inquiry in universes of the future, when people will look back and no doubt also see our instruments as inappropriate.) The phenomena of demonology and the facts of witchcraft were repeatedly tested, were indeed falsifiable, and many inquisitorial experimenters and theorists even hoped that the facts would be falsified. But the facts stood out stark and clear, plain for all to see, inalterable and fully verified. "Would to God," said Kramer and Sprenger, "we might suppose that all this to be untrue and merely imaginary, if only our Holy Mother the Church were free from the leprosy of such abominations."

In pursuit of this theme, imagine that we are precipitated into a universe where numerous men and women associate with demons. We sally forth in a spirit of inquiry and discover that the history of this universe provides testimony and provenance in full support of the fact that many of its occupants consort with demons. We find that all institutions sanction this fact, everybody is fully convinced of its truth, and wizards and witches themselves give chapter and verse, albeit under duress, recounting their experiences with demons. After becoming inhabitants of this universe, we form the belief that here, indeed, is a universe containing people who associate with demons, and this fact, though constantly exposed to the perils of falsification, remains incontrovertible.

The witch universe was falsifiable. By Popper's definition of what constitutes science, the witch universe was therefore scientific. This conclusion is patently false by our modern standards. Though we do not know exactly what is science, we must all agree that the witch universe was not scientific by any stretch of the imagination. The falsification theory of what constitutes science is therefore itself falsified. A theory that condemns astronomy and biological evolution as unscientific, yet accepts the witch universe as scientific, has missed the point.

□ □ □

The internal relations of a universe determine what are falsifiable facts, and the falsifiable facts support the relations. Falsifiability applies in all universes, affirming each particular universe, rejecting as unreal what is unfalsifiable and therefore not an integral part of that universe.

Falsification operates in all universes, it props up the "true" facts as perceived, deals only with relative truths and not with absolute truths, and is quite unable to distinguish between the merits of different universes in any absolute sense.

The falsification argument applies equally to the witch universe and the modern physical universe. Furthermore, in whatever universe we find ourselves, the supporting facts are at hand, falsifiable, yet repeatedly verified.

We must agree with Popper's theory that science consists of what is falsifiable, subject to two conditions. First, we permit the falsifiable to include chains of reasoning that link together whatever is observed, as in astronomy and biological evolution. Second, we recognize that the theory fails to tell us what is science and is poverty-stricken to the point where it cannot distinguish between science and witchcraft.

The occupants of any universe, whether magic, mythic, geometric, or mechanistic, tend to regard their universe as falsifiable within its own terms of reference, in accordance with its own criteria of what constitutes valid knowledge. In the physical universe we regard science as falsifiable, and we use falsification as the demarcation between acceptable and unacceptable knowledge; in the mythic universe we regard the gods and their works as falsifiable, and again we use falsification as the demarcation between acceptable and unacceptable knowledge. The conclusion we must draw is that falsification distinguishes between the things that fit naturally and the things that fit unnaturally in the universe we happen to occupy.

We find in any universe that what is fitting is falsifiable, and what is unfitting is either falsified or unfalsifiable. In the modern physical universe the myth that angels propel the planets is falsified, and the myth that angels exist is unfalsifiable. According to our definition of myth (see *The Mythic Universe*), all vestiges of previous universes are mythic if either falsified or unfalsifiable. Always, in any universe, the falsified and unfalsifiable inherited elements of knowledge are mythic.

Hypotheses as provisional truths are freely invented, slapped on here and there like the bits of clay used by a sculptor, then examined and scraped off when they fail to conform to the desired shape. But where is the master diagram that determines the shape to which the clay must conform? The diagram in the minds of the world shakers and makers, the diagram that determines what is fit and unfit, is itself transitional, doomed to be scrapped as mythic and replaced by the diagram of a new universe.

The notion that attested knowledge consists of facts, verified until falsified, works tolerably well in fields of inquiry capable of defining what is fit and unfit, what is true and untrue. It applies to science in the physical universe, witchcraft in the witch universe, and salvation in the medieval universe. The facts of attested knowledge, though vastly different in each, stand moderately secure.

Like the sculptor's lump of clay, a universe is shaped according to a diagram. But the diagram changes, and the universe evolves. New facts and ideas intrude, meanings and relations become modified, leading to the discovery of more facts and ideas. Continually a universe changes, the falsifiable becomes falsified or unfalsifiable, and continually a new universe is in the making.

What the next universe will be like is never known, but within its consistent and complex scheme of falsifiable knowledge the old "facts" will be discredited and either reinterpreted or discarded.

14

The Spear of Archytas

The Universe is everything, including the universes of societies and the world pictures of individuals. Each universe unifies a society and dictates the "true" facts. Individuals suppose with unfailing optimism that their universe is the Universe, and feel quite unconcerned that it was greatly different in the past and might be greatly different in the future.

The impersonal universe of a society is always vast, rational, and deterministic. The personal world pictures of individuals are always muddled, irrational, and emotional. Individuals believe they have the liberty to act and choose as they wish and yet at the same time believe in the determinism of their societal universe.

We feel at home in our world pictures but lost in the universe. We depend on wise men—shamans, priests, sages, prophets, and professors—to put us right and tell us the "true" facts. As long as somebody knows the truth, all is right with the universe.

To every universe we may ascribe a containment principle that more or less defines what is included as fitting and excluded as unfitting. A containment principle is a thumbnail sketch that separates the mythic (falsified and unfalsifiable) from the real (verified and falsifiable). Thales said the Ionian universe consists of water; Anaximenes said air; Heraclitus said fire; Xenophanes said earth; Empedocles said earth, water, air, and fire. Democritus said the atomist universe consists of atoms and the void, and all else is illusion and opinion. Plato said the Platonic universe consists of the eternal verities of the Mind, and all else is shadow and deception. Aristotle said the Aristotelian universe consists of earth, water, air, fire, and ether in ascending order, animated by Ideas, with nothing beyond the sphere of the stars. Saint Augustine said the Christian universe consists of the Word of God, and all else is heresy. John Wheeler said the physical universe consists of "empty curved space," and "matter, charge, electromagnetic and other fields are manifestations of the curvature of space."

In every age people believe that their universe contains whatever is real and significant. In their temples, academies, monasteries, and universities they reject the rest as opinion and illusion. Forget the superstitions of the uneducated and the myths your parents taught you. For behold! Here is the true universe, awesome, vast, and wondrous: the world is a tug-of-war, with gods and demons pulling on a giant serpent; the world is the handiwork of almighty gods whom we must obey and worship; the world is a geocentric unity of crystalline spheres; the world is a dance of atoms and waves—all else is myth and fancy. The scene is timeless. Yesterday a false universe, today the true universe.

I suspect that containment principles are necessary because of the difference between universes and world pictures. The more primitive a society the less the difference between the impersonal universe and its personal world pictures, and the less the need for containment principles. The more complex a society, the greater the difference and the greater the need. Rules of containment protect the universe from being swamped by the errant fancies of its diverse world pictures. Without such rules, defining what is rational and fitting, rejecting what is irrational and unfitting, the universe would collapse and society break down.

□   □   □

The physical universe contains all that is physical and nothing else. This is the containment principle of the physical universe.

Atoms and cells, flowers and planets, stars and galaxies are physical things studied by the natural sciences. Atoms and their electron waves, replicating organisms and their differential survival, evolving stars and their transmutation of elements, space and time and their dynamic warping belong to the physical universe that contains physical things and nothing else. DNA molecules with their genetic coding, our bodies born and doomed to die, our brains throbbing with bioelectronic activity are the stuff of the physical universe.

But the thoughts and the mental furnishings of our world pictures are unphysical, hence uncontained and therefore "unreal" except for analogous forms of cerebral activity and glandular chemistry.

Learned savants assure us that our modern universe is the Universe, that all happenings have their physical causes, all objects consist of molecules, atoms, and subatomic particles, and what is unphysical does not exist in the physical universe. They teach us that organic life evolved over billions of years on Earth, that our spinning Earth orbits the Sun, that our shining Sun is a star far from the center of the Galaxy, that our whirling Galaxy of stars is but one in a vast universe of galaxies, and that the universe expands.

The physical universe contains all that is physical and nothing else. This is an eminently sensible principle, provided we do not inquire too closely into the meaning of *physical*. The physical universe changes, and the meaning of physical changes with it. The ponderable matter of Dr. Johnson's stone has lost its concreteness in a cloud of probability waves.

The first message of the containment principle of the modern world says that space and time are physically real—real as apple pie—and spacetime is a dynamic constituent of the universe. Space and time are contained and therefore cannot extend beyond the universe. As we shall see, this has important consequences.

The second message, more difficult to understand, says that images do not contain the image maker. Our universes do not contain world pictures, in other words, do not contain us who create universes. Each universe arises as a product of the mind and contains, at best, only a representation of the mind that creates it. Credit for this remarkable discovery belongs to the physical universe and its clearly enunciated containment principle.

Containment causes much perplexity. We agonize over freedom of will, moral responsibility, and the enigma of how nerve impulses translate into thoughts. Our anguish is needless. Freedom of will and the mind with its thoughts belong to the Universe; determinism and the brain with its nerve impulses belong to the universe. Recognition that the physically uncontained may actually exist in the Universe, though not in a universe, cuts like a bracing wind through a world of discourse.

Space and time are physically real, and therefore, according to the modern containment principle, are contained in the physical universe. By journeying in space and time we are unable to escape from the universe. The ancient view, popular until recent times, held that space and time contained the universe. The modern view is the opposite: the universe contains space and time.

Cosmic edges defining the limits of the universe were once of great importance in cosmology. A cosmic edge implied a cosmic center, and both—edge and center—propped up the anthropocentric universe. The center was first the nation, then the Earth, the Sun, and finally the Galaxy. The edge, however, was much less obvious and posed problems that taxed the ablest minds.

What happens when a spear is thrown across the cosmic edge? This is the famous cosmic-edge riddle, known in the Middle Ages. The riddle can be traced back to Archytas of Tarentum, the Pythagorean philosopher-scientist, statesman, soldier, friend of Plato, and possibly the model for Plato's philosopher-king. The spear of Archytas shatters universes and opens the way to the study of containment.

The Roman poet Lucretius used the riddle of Archytas to great effect:

> Suppose for a moment that the whole of space were bounded and that someone made his way to the outermost boundary and threw a flying dart. Do you choose to suppose that the missile, hurled with might and main, would speed along the course on which it was aimed? . . . With this argument I will pursue you. Wherever you may place the ultimate limit of things, I will ask you, "Well, then, what does happen to the flying dart?"

Does the spear rebound, continue on its way, or disappear? Lucretius gave the atomist answer: "Learn, therefore, that the universe is not bounded in any direction."

Possibly Lucretius's epic poem *The Nature of the Universe*, and hence the spear of Archytas, influenced Cardinal Nicholas of Cusa, who saw in the infinity and ubiquity of God the justification for a centerless and edgeless universe. Possibly the poem and the spear influenced also the astronomer Thomas Digges, who tore away the outer edge of the Copernican system, and the physician William Gilbert, who advocated the infinite atomist universe of numberless celestial worlds, and also the luckless Giordano Bruno. In his *Infinite Universe* Bruno wrote, "If a person would stretch out his hand beyond the convex sphere of heaven, the hand would occupy no position in space, nor any space, and in consequence would not exist. . . . Thus, let the surface be what it will, I must always put the question: what is beyond?" Only one answer seemed to resolve the riddle: the universe was edgeless, hence limitless and infinite in size.

The cosmic edge. (E. R. Harrison, *Cosmology: The Science of the Universe*, Cambridge University Press, New York 1981.)

We can identify three kinds of cosmic edges. First and most ancient is the idea that the universe ends abruptly in a wall-like edge. The universe is like an egg bounded by a shell or a cave bounded by walls. The answer to Archytas's riddle is that the thrown spear rebounds and remains inside the universe. Lucretius, echoing the atomists, rejected this answer by arguing that space is continuous and cannot end: "It is a matter of observation that one thing is limited by another. The hills are demarcated by air, and air by hills. Land sets bounds to seas, and seas to every land. But the universe has nothing outside to limit it." Johannes Kepler, who abhorred the notion of a boundless universe, was not convinced. At night, said Kepler, when we look out between the stars we see a dark enclosing wall; the darkness of the night sky is nothing more than the darkness of the enclosing wall. Imagine what would befall us if the universe stretched away endlessly, populated with numberless stars, as the ancients claimed. Every line of sight would eventually terminate at the surface of a distant star, every point of the night sky would blaze with starlight, and the whole sky would be as bright as the Sun.

Why is the night sky dark in a boundless universe? This riddle, which derives directly from Archytas's cosmic-edge riddle, has taken cosmologists many centuries to solve. We know now that when we look out to vast distances, we also look back to a time before the birth of the first stars. We cannot look out to unlimited distances in a universe of finite age, and for this reason the number of observable stars is insufficient to cover the whole sky. Independently of whether the universe expands or is static, whether it consists of stars distributed uniformly or clustered into galaxies, or whether it is open or closed, we see empty spaces in the night sky. This is why the sky is not ablaze at every point with starlight.

The second is the clifflike edge adopted by the Stoics of the Greco-Roman world. In the Stoic universe the stars formed a finite spherical cosmos, and beyond the edge of the cosmos extended a void of unlimited empty space. The spear crossed the edge of the cosmos and was lost in the void. The void necessarily existed in order to solve the riddle of Archytas.

The third is the Aristotelian edge. In the Aristotelian and medieval universes, physical things existed only in the sublunar sphere. If the spear passed beyond the lunar sphere, it necessarily became etheric, and its natural and only possible motion was then circular around the Earth. The physical realm merged into the etheric realm, and an abrupt boundary between the two did not exist. The thrust of "with this argument I will pursue you" was lost, because the pursuer was led into a metaphysical realm where the argument lacked validity. The Aristotelian edge faded gradually, like the fading of firm ground on entering a marshland.

The physical universe is not like an egg or a cave. Space always continues and never ends abruptly at a wall-like edge. Whatever lies "outside" must still belong to the universe.

The compromise Stoic or clifflike edge was not so easily dismissed. William Herschel adopted the Stoic picture in the eighteenth century, and in the early decades of this century the idea of an island universe— our Galaxy surrounded by an endless void—gained support. Harlow Shapley, a famous astronomer in the first half of the twentieth century, tried to resolve Kepler's dark-sky riddle by adopting a Stoic cosmos. Astronomers have since discovered that the universe of galaxies extends to enormous distances without any sign of a clifflike edge.

The Aristotelian or marshlike edge also lingered on. I remember the teacher in a scripture lesson pointing to the ceiling, when asked where is God, and saying, "Up there in Heaven." More than 2000 years previously Anaxagoras had said the constituents and laws of the universe are the same everywhere; only a decade previously Einstein had shown with the theory of general relativity that space and time are physically real. Whatever occupies physical space must itself be physical, and therefore, a nonphysical realm cannot exist within the space and time of the modern universe.

Beginners in modern cosmology often have in mind a simple picture resembling the Stoic cosmos. They suppose that space extends into a void from which it is possible to view the expanding universe. From this imaginary extracosmic grandstand they see the universe as an exploding swarm of galaxies, having an expanding outer edge, with the big bang at the center. But this picture is misleading, because no outside void exists from which the universe can be observed. The big bang did not occur at a point in space, for it occupied the whole of space in the early universe.

Cosmic edges have gone, and with them the cosmic center. Space is unbounded, and it may be infinite in extent, or finite as on the surface of a sphere.

☐　☐　☐

A miscellany of subjects falls under the heading of containment. Some subjects are elementary. Thus contained things can be neither larger nor older than the universe; a galaxy, for example, is necessarily smaller and younger than the universe. Other subjects, such as the nature of mind and free will in a deterministic universe, are not in the least elementary. First we look at the subject of cosmogenesis, now transformed by twentieth-century ideas of containment.

Creation for our purpose has three distinct meanings.

Scientists use the word creation to denote a change in which something transforms, and in the process fundamental quantities remain conserved. Thus the creation of a particle and its antiparticle conserves energy and electric charge.

Next is the miraculous kind of creation in which something is created from nothing. At a place in space at a moment in time there is nothing, and at the same place a moment later there is something from nowhere.

Then there is the third kind in which "God created the heaven and earth."

Creation is cosmogonic, miraculous, or cosmogenic. Cosmogony deals with the formation and evolution of the components of the universe and is the only kind of creation recognized in the sciences. The origin and evolution of life on Earth is cosmogonic. Miraculous creation generally deals with mythical matters. Cosmogeny, or cosmogenesis, deals with the creation of the universe. Scientists will have nothing to do with miracle making; and most scientists shy away from cosmogenesis, believing it to be metaphysical and mythical and not a fit subject for scientific inquiry.

Cosmogony and miracle making deal with creation within the universe, whereas cosmogenesis deals with the creation of the whole universe. Thus we have two basic kinds of creation. The first (cosmogony and miracle making) refer to contained creation, and the second (cosmogenesis) refers to uncontained creation. By this I mean the first kind is creation within the fabric of space and time, and the second kind is the creation of the whole universe that includes the fabric of space and time. Cosmogenesis necessarily includes the creation of space and time, because both are physical, like everything else in the physical universe.

Hermann Bondi, Thomas Gold, and Fred Hoyle in 1948 proposed a steady-state universe in which matter is continuously created everywhere. They discarded conservation of energy and replaced it with what seemed to them more fundamental: the conservation of the universe in its present form. In the expanding universe the widening gulfs of space between old galaxies become occupied by young galaxies born from newly created matter, thus preserving the appearance of the universe. In the controversy following this proposal many contestants held the view that the instant creation of a big-bang universe is of the same kind as the continuous creation of a steady-state universe. We have the choice, it was argued, of cosmogenesis all at once or little by little. But the creation of a whole universe is very different from the creation of its bits and pieces; the former is uncontained and the latter contained within space and time.

Cosmogony and miracle making deal with creation in the space and time of an existing universe; cosmogenesis deals with creation of the whole universe including its space and time. Failure to distinguish between contained and uncontained creation violates the containment principle of the physical universe. Cosmogenesis cannot have the same meaning as miracle making unless we revert to the mistaken view that the universe is contained in a preexisting space and time. Nowadays we are not free to say that at a place in space there is nothing and an instant later at the same place there is a universe.

Some cosmologists have strongly disliked the idea of cosmic birth and death. The most notable was Arthur Eddington, who in 1930 constructed a model universe having an infinite past and an infinite future. With his eternal universe he sought to escape the mythological specter of creation. He failed because cosmogenesis is the creation of a whole universe that comes endowed with time. The physical universe is not created in time, and cosmogenesis (independently of whether it actually occurs) cannot in principle be pushed out of sight into an infinite past. A universe with an infinite span of time is created just the same as a universe with a finite span. From a cosmogenic viewpoint the eternal Eddingtonian and steady-state universes are the same as any other. All universes of finite or infinite duration in time, finite or infinite extension in space, with or without big bangs are confronted with the possibility of cosmogenesis.

We must realize that the physical universe is not created somewhere in space at a moment in time. The big-bang universe is not created in the big bang. A physical universe is created in its entirety, equipped with a vast expanse of spacetime that includes the beginning and the end. To think otherwise violates the containment principle of the physical universe and preserves falsified mythic beliefs.

□　　□　　□

The mythic universe was created by God about 6000 years ago according to Mosaic chronology. Trouble occurs when the mythic universe is confused with the physical universe. For the moment, let us forget the estimated age of the physical universe. It is the act of creation claimed by mythology to have occurred at a particular moment in time with which we are interested.

The special creation theory holds that life is miraculously created in a world already existing and previously created. Special creation is therefore an act of contained creation. Let us now suppose that special creation occurs in the physical universe. Clearly, if the whole universe is first created as a continuum of spacetime, then everything in the universe, including the origin of life, is already present and arrayed in space and time. What is created once has no need to be created twice. In a created universe acts of theistic intervention are superfluous, for all is already created and decreed. Thus cosmogenesis preempts the miraculous creation of life. Special creation may occur in the mythic universe but not in the physical universe where it violates containment.

Perhaps this point should be made more clear. Let us assume that God creates the physical universe. Further acts of creation are unnecessary and even illogical. The universe is created in neither space nor time, because space and time are physical and created with the universe. The created universe is complete in every detail throughout all of space and time. Subsequent acts of creation meddle with what is already created, implying the possibility that the Creator actually exists in physical space and time, and to this extent is therefore a physical and not a spiritual being.

Although not a science, cosmogenesis nonetheless is constrained by the logic of containment. Knowing that the universe cannot be created in space and time, because it contains both, we must draw the conclusion that the special creation of life in the physical universe is falsified and mythic.

There is a further consequence. No useful purpose is served in the physical universe by praying for theistic intervention, such as the award of victory in battle, for the events of the past and future are created together, and all are inflexibly ordained. Some persons may not like the physical universe, and in that case, of course, there is nothing to prevent them from enjoying the comfort of the mythic elements in their world pictures. But in no circumstances may they attribute these elements to the physical universe.

"God could create and without doubt did create the world with all the marks that we see of old age," said Chateaubriand, the French deist. Many deists reconciled the scriptural records with the physical evidence by supposing that God created a world already bearing the signs of great age. The zoologist Philip Gosse also saw no reason for abandoning the Mosaic chronology. In his book *Omphalos* (a word meaning navel), published in 1867, Gosse argued that Adam was created with a navel and carried the vestiges of a birth that had never occurred. If God saw fit to create Adam with a navel, said Gosse, then surely God also saw fit to create a world equipped with the vestiges of past eras that had in fact never occurred. The universe was created some thousands of years ago and outfitted in that supreme cosmogenic act with a fictitious history of hundreds of millions of years. The impact of raindrops etched in sedimentary clays, the footprints and teethmarks of primordial beasts, the fossils embedded deep underground, the light in transit from distant stars, and all the elaborate ornamentation of a complex and consistent universe bearing the signs of great age were created in the recent past.

In the same spirit we might argue that the universe was created this morning before breakfast. The world comes equipped with a spurious past and is inhabited by people with false memories, as in a historical novel. This is an unfalsifiable argument and therefore unscientific. The Mosaic chronology, when applied to the physical universe, makes creation a hoax and God a jester.

Who created the physical universe? The answer consistent with the theme of this book is that we did. The Universe contains us who created the physical universe, and the physical universe is our attempt to understand the Universe.

Then who created the Universe? The question is meaningless, because the Universe is inconceivable. If a person finds comfort in the notion that God created the Universe, then so be it; there is nothing to gainsay this unfalsifiable belief. But those persons who hold that God created the physical universe, and subsequently meddled with it in comformity with mythology, violate containment by imposing the irrational elements of their world pictures on the rational universe of their society.

◻   ◻   ◻

In *Our Mutual Friend* Charles Dickens wrote, "Mr. Podsnap settled that whatever he put behind him he put out of existence." He "had even acquired a peculiar flourish of his right arm in often clearing the world of its most difficult problems by sweeping them behind him." The Podsnap flourish is encountered in all walks of life, and whatever perplexes is conveniently swept out of existence. Many persons believe that what is not contained and explained by the physical universe does not exist.

What is the nature of mind in the physical universe? The response is often the Podsnap flourish. Psychologists know much about mental things, but not about physical things, and their response might be: Why not ask the physicists who deal with strange things like electron waves and what they call the uncertainty principle? The physicists, when asked, respond by saying: The brain is certainly part of the physical world, but as for the mind, well . . . why not go and see the psychologists? You have already spoken to them? Then try the biologists, they know a lot about organic things. The biologists, in their turn, respond by recommending a visit to the physicists: You have already spoken to the physicists? Well why not try the psychologists?

Wherever inquiry is made concerning the nature of mind in the physical universe the inquirer either encounters the Podsnap flourish or is passed from specialist to specialist, like a patient with a strange disease.

Cosmologists—those who try to look at things in general—are denied the luxury of the casual Podsnap flourish. Their task is to assemble the jigsaw puzzle of the universe, and they are not at liberty to throw away awkward pieces willy-nilly without justification. Confronted with the burning question—What is the nature of mind in the physical universe?—they do what cosmologists have always done with awkward questions. They enunciate a limiting principle of containment equivalent to a glorified Podsnap flourish. The containment principle of modern times asserts that the physical universe contains physical things and nothing else. The inquirer is told that the brain is included in the physical universe, but the mind, if it exists, must be found elsewhere.

Theologians, philosophers, and scientists—those of them who are professional cosmologists—realize perhaps more than anybody else that the universe is a model. They are not greatly troubled about excluding something that fails to fit naturally into the rational scheme of things. Inside their temples, ivory towers, and laboratories they rarely take the universe as seriously as the rest of the population. In their bones they know that the universe falls a long way short of the Universe.

The inquirer might easily think that the physical universe has let us all down, and the answer must be found in a Taoist, Buddhist, or some other universe. The question addressed to the cosmologists now becomes quite general: What is the nature of the human mind in other universes? We are at last face to face with the problem of containing the human mind in any universe, and not just the physical universe that has brought the problem to the surface.

Containment poses two riddles. The first, relating to cosmic edges, has been solved by the containment of space and time within the universe. The second is what I call the containment riddle. The containment riddle asks, *Where in a universe is the cosmologist conceiving that universe?* By cosmologist I mean anybody who stands back and looks at things in general. The riddle applies to all universes—Zoroastrian, atomist, Aristotelian, medieval, Newtonian—and not just the modern physical universe.

The riddle is made clear by using the analogy of a painter who paints a picture of the interior of a studio. A complete and faithful picture must show the studio containing the painter in the act of painting the picture. But this entails an infinite regress: the picture contains the painter who paints the picture that contains the painter who paints the picture that contains . . . and so on indefinitely.

Cosmology is the art of creating cosmic pictures or universes. If we claim that a universe is a complete and faithful representation, then it must contain the cosmologist conceiving and thinking about that universe. This also entails an infinite regress: the universe contains the cosmologist conceiving the universe that contains the cosmologist conceiving the universe that contains . . . and so on indefinitely. Unless the mind that conceives and studies a universe is excluded, we lapse into infinite regress.

The *Tale of a Danish Student* by Poul Møller is an amusing account of a soul-searching student perplexed by the intricacies of human thought. The student thinks,

> . . . man divides himself into two persons, one of whom tries to fool the other, while a third one, who in fact is the same as the other two, is filled with wonder at this confusion. In short, thinking becomes dramatic and quietly acts the most complicated plots with itself and for itself; and the spectator again and again becomes actor.

But for the puzzled student thinking becomes hectic,

> . . . and then I come to think of my thinking about it; again I think that I think of my thinking about it, and divide myself into an infinitely retreating succession of egos observing each other. I don't know which ego is the real one to stop at, for as soon as I stop at any one of them, it is another ego again that stops at it. My head gets all in a whirl with dizziness, as if I were peering down a bottomless chasm, and the end of my thinking is a horrible headache.

Niels Bohr was much impressed in the early days of quantum mechanics by this amusing tale. Leon Rosenfeld remarks, "Poul Møller had little imagined that his lighthearted banter would one day start a train of thought leading to the elucidation of the most fundamental aspects of atomic theory and the renovation of science." The epistemic problem of what is observing what in the physical universe lies at the heart of quantum theory and the philosophy of science.

On a similar subject Erwin Schrödinger in *Mind and Matter* writes,

> Sometimes a painter introduces into his large picture, or a poet into his long poem, an unpretending subordinate character who is himself. Thus the poet of the *Oddysey* has, I suppose, meant himself by the blind bard who in the hall of the Phaeacians sings about the battles of Troy and moves the battered hero to tears. In the same way we meet in the song of the Nibelungs, when they traverse the Austrian lands, with a poet who is suspected to be the author of the whole epic. In Dürer's *All-Saints* picture two circles of believers are gathered in prayer around the Trinity high up in the skies, a circle of the blessed above, and a circle of humans on the earth. Among the latter are kings and emperors and popes, but also, if I am not mistaken, the portrait of the artist himself, as a humble side-figure that might as well be missing. To me this seems to be the best simile of the bewildering double role of mind.

Reductionists and materialists, overwhelmed a little by the power and glory of the physical universe, are inclined to suppose that what is not contained does not exist. Mind and free will are dismissed as illusions. Those who adopt this Podsnap flourish must show us where they themselves are fully portrayed in the physical universe in the act of conceiving that universe in which mind is an illusion.

It is of little use pointing to the brain, because this leads also to an infinite regress: the brain conceives the universe that contains the brain conceiving the universe that contains the brain . . . and so on indefinitely. How can a thing conceive a whole of which it is a part conceiving the whole of which it is a part conceiving the whole of which . . . ? Whatever conceives the universe is not in the universe and cannot therefore be just the brain.

We cannot point to the physical brain, with its networks and discharge patterns, and say here is the mind. For in its cerebral activity lies a representation of cerebral activity that would be us saying the mind is the brain. The interplay of neurons conceives the interplay of neurons that conceives . . . and we are on the slippery slope of an endless regress. Nowhere in the interplay can be found the world of mental experience, and therefore, by means of an eminently sensible containment principle, the undeniable mental world is conveniently and provisionally excluded from the physical universe.

Vitalists less content with the physical universe seek to enliven its organic structures with a psychic efflorescence possessing the virtue of "life." The intricate dance of multitudinous atoms aquires a spiritual aura. This animistic theory explains nothing, and furthermore, it violates modern containment. The vitalists must show us where in their vitalized universe they themselves are fully portrayed in the act of conceiving a vitalized universe in which the mind is a product of physical systems. Needless to say, they also run into an infinite regress.

The containment riddle applies to all universes, ancient and modern. Even the theologian conceiving a theocosmos must show us where in the theocosmos is the theologian conceiving the theocosmos. All who claim to know what is mind within their particular universe make the mistake of confusing a universe with the Universe. It is the conceit of supposing that we have plumbed the depths and discovered the Universe, the conceit of thinking that a universe contains us understanding the inconceivable Universe.

All who think their universe is the all-embracing Universe are obliged to answer the riddle. In my opinion the riddle is unanswerable, because only the inconceivable Universe contains minds that conceive universes.

Thomas Hobbes, born at the time of the invasion of England by the Spanish Armada, wrote in the *Leviathan*, "for what is the heart but a spring, and the nerves but so many strings, and the joints but so many wheels giving motion to the whole body." René Descartes followed this avenue of thought to its conclusion, and to him we attribute the total mechanization of the external organic and inorganic world. "I think, therefore I am," he said, but everything else in the objective world consisted of configurations of matter in motion. Give him matter and motion, and he would construct the universe. Human bodies, with their brains and sense organs, obeyed the laws of the clockwork universe.

The mind was a ghost haunting the machinery of body and brain. Thus began the notorious Cartesian duality of mind and matter that proliferates paradoxes whenever any attempt is made to put the mind into the physical universe. Descartes discovered reassuring evidence of his own mental existence; he then assumed that the mechanistic universe was the Universe and found himself confronted by a paradox: mind and matter manifestly interacted, yet logically were incapable of interacting.

We exist, and the Universe of which we are a part or some aspect also exists. But, and here we must part company with Descartes, the universe conceived by us is not the Universe and cannot contain us. Instead of doubting his own existence, Descartes should have doubted the universe he conceived and took for granted.

Erwin Schrödinger in *Mind and Matter* wrote, "I so to speak put my own sentient self (which has constructed this world as a mental product) back into it—with the pandemonium of disasterous logical consequences." And Alexandre Koyré in *Newtonian Studies* added, "This is the tragedy of the modern mind which 'solved the riddle of the universe,' but only to replace it by another: the riddle of itself." The "riddle of itself" boils down to the containment riddle.

The brain throbbing with ceaseless electrochemical activity is part of the physical world, and though much is not understood about the brain, steady progress is being made. The brain and the computer have much in common, and developments in computer science undoubtedly will help us to understand more about the brain. Through their input channels the brain and computer receive information that is modified, processed, and stored, and both respond in elaborate ways to their environments. It does not take much imagination on our part, only a little courage, to foresee a time when artificial intelligence will perhaps rival and then outrival human intelligence. All this within the context of the physical universe.

Human beings are by no means perfect, and with their muddled and emotional thinking, they seem incapable of making clear-sighted decisions on which our survival depends. The prospect of surrendering control of our political and judicial systems to pondering logical machines of advanced artificial intelligence does not bother me; sometimes I think they are the one ray of hope for a sane future. Modern society is far too complex for the human mind to compass and regulate in a sane and sensible fashion. The master machine-minds of America and Russia, conferring together, would realize in a microsecond that nuclear weaponry is not in the best interests of their human populations and in a millisecond would work out how to abandon these weapons forthwith.

The notion of artificial intelligence outrivaling human intelligence is abhorrent to most persons. These slavelike computing creatures of the physical world, which one day will be self-adjusting and self-programming, must at all costs be kept in their proper places. We hide behind an assumed monopoly of intelligence and protest when anyone attempts to define or measure intelligence. We soothe our fears by talking about the Turing Test and the impossibility of inhuman machines outmatching human brains.

I am tempted to think that whatever the brain can do the machine will one day do better. Many of our fears of artificial intelligence stem from the belief that the physical universe is the Universe, and therefore we must preserve our status as superior physical entities. There is no need for fear; we are the creators of the physical universe in which artificial intelligence exists, yet we are ourselves a part or an aspect of the unknown Universe.

□　　□　　□

There remains the vexed subject of individual freedom of will versus determinism, which applies not only to the physical universe but to all universes.

A universe makes rational the experiences of individuals who live in it. Whatever befalls those individuals has its natural explanation, and the explanation is peculiar to that universe. The world is an activity of spirits; the world is dead matter jerked into motion by strings in the hands of gods; the world is a vast clockwork mechanism, and the human soul belongs to God; the world is a dance of atoms and waves observed and studied by electrochemical brains. In each case the world is lucid, rational, and deterministic.

Human beings exercise will and have an awareness of personal freedom. Yet as inhabitants of a rational universe they are components of that universe governed by its deterministic laws. They plan, but the whim of spirits controls what happens; they believe they have the freedom to do either this or that, but in fact the fiat of gods determines everything; when overwhelmed by events, they are comforted by the thought that

God knows best; they have the illusion of liberty to go either here or there, but know that all was fated long ago when the machinery started on its predestinate grooves. They acquire wealth rather than live in poverty; climb mountains and go to wars rather than stay at home; join protest movements rather than rest content; study rather than watch television; all the time automatically assuming deep down that the choice is theirs, yet knowing they are at the mercy of the dancing atoms, are caught in the coils of their genetic coding, and must follow a destiny shaped by their inheritance and environment.

We live in our rich world pictures of muddled fantasy, and all is redeemed and made clear by the simple lucidity of the universe of our society. But the price of a universe—any universe—is that freedom of will becomes an illusion of our world pictures.

Marcus Aurelius, a Roman emperor and Stoic, wrote in his *Meditations*, "Whatever may happen to you was prepared for you from all eternity; and the implication of causes was from eternity spinning the thread of your being." Augustine, the Christian, in his *Confessions* stressed the logical necessity of predestination in a created universe. The Stoics believed in fate, and the early Christians believed in predestination. The contrary belief that human beings have freedom of choice is the essence of the Pelagian heresy that mocks all universes. Pelagius, a British theologian and monk of the early fifth century, was appalled by the decadence of social life in Rome. His protests met only excuses and the evasive plea that human weakness had been preordained by God. Wickedness being inevitable is hence forgiveable, he was told.

The prevailing doctrine, elaborated by Augustine and now at the heart of Christianity, declared that everything was predetermined by the will of almighty God. This was not good enough for the down-to-earth Pelagius, who lacked Augustine's conviction that the ways of God are transparent to rational inquiry. A predestinate universe threw on God the blame for wickedness that properly belonged to human beings.

Pelagius believed in freedom of individual will. He argued that the grace of God must be earned by righteous living and is a gift to all and not just a few. He took a view contrary to that of Jerome and Augustine, and he taught that God had not created an inalterable world of good and evil. Men and women had the freedom, if they so willed, to live untainted by sin, inasmuch as God had given them that freedom. Original sin did not exist. "If it is necessary, then it is not a sin," he said, "if it is optional, then it can be avoided."

There was much dispute, and Augustine won the battle with the aid of scriptural testimony. A rational universe, predetermined and controlled by divine will, in which human beings lacked free will and behaved as robotic creatures, triumphed. The Pelagian defense of human free will was condemned as heresy in 418.

Pelagius perceived that the rational universe of his time was not the Universe. He in effect excluded the mind (or the soul) from the universe and retained freedom of human will. This is the essence of the Pelagian heresy. In every age orthodox believers dismiss free will as an illusion, for they believe that their universe is the Universe.

Freedom of will belongs to the Universe, determinism to a universe, and unto Caesar we render only that which is Caesar's.

15

All That Is Made

In the tenth and eleventh centuries ibn al-kalam (a religious science) opposed the Aristotelian philosophy of Muslim theology. The dialecticians of the kalam (known as the mutakallimun) adopted the atomic theory of the Epicureans and professed a theory of extreme theism. The universe, they said, has no natural laws; instead, it is governed by the law of the Sole Agent. Al-Bakillani, a disciple of a disciple of the famed al-Ashari (the founder of Muslim scholasticism), lived in Baghdad and died in 1013; he introduced an atomic doctrine that included atomic time.

The anti-Aristotelian philosophy of the mutakallimun was critically discussed in *The Guide for the Perplexed* written in the twelfth century by Moses Maimonides. The kalam atoms, as described by Maimonides, were noninteracting and much like the monads of Gottfried Leibniz. (Leibniz possessed an annotated copy of Maimonides's *Guide* and was aware of the kalam atomic theory.) Not only matter but also space was finely divided in the kalam universe; nothing interacted, and nothing bridged the atomic gulfs except the harmonizing power of the Sole Agent. The mutakallimun reconciled the discontinuity of the atomic world with the continuity of the macroscopic world by assuming that all motion occurred in imperceptibly small leaps and jerks.

But if nothing interacts, everything is discrete and isolated, and motion always discontinuous, how does the transcendental principle of divine interference account for the coherence and order apparent in the universe? Al-Bakillani argued that time must also be atomic. "An hour," explained Maimonides, "is divided into sixty minutes, the minute into sixty seconds, the second into sixty parts, and so on; at last after ten or more successive divisions by sixty, time-elements are obtained, which are not subjected to division, and in fact are indivisible, just as is the case with space."

The universe according to the mutakallimun throbbed with recurrent cosmogenesis. In every atom of time the kalam universe was decomposed and recomposed by the Creator. In a single time-atom the recreated universe unfolded in a static state of being, then dissolved, and in another time-atom again unfolded in a new and different static state of being. No intervals of time separated the time-atoms, and the passage of time was inferred by comparing the states of being in the different time-atoms.

The kalam atomic universe consisted of cosmic states of being, each decomposed and recomposed by cosmic acts of becoming. In each atomic state of being everything was accidental and incidental; in each act of becoming everything was theistically determined and coordinated.

Not improbably the kalam atomic theory of time originated with the pre-Socratic Greeks. We shall never know. About 1 percent, perhaps less, of the literature of the ancient world has survived.

Each age writes its own history and is convinced of the authenticity of that history. Each age adjusts the records so that others who follow shall not be misled by discordant facts. In other words, it erases either deliberately or by neglect the literature, works of art, artifacts, and architecture of the past. The libraries of Alexandria and Baghdad were burned; the language and much of the Mayan culture obliterated; the works of Copernicus and Galileo placed on the Index Expurgatorius. It is the human arrogance that in each age thinks its history is History, its universe is the Universe.

□　　□　　□

Julian of Norwich in the fourteenth century wrote in her *Revelations of Divine Love*, "I saw that he is everything that we know to be good and helpful."

> And he showed me more, a little thing, the size of a hazelnut, on the palm of my hand, round like a ball. I looked at it thoughtfully and wondered, "What is this?" And the answer came, "It is all that is made." I marveled that it continued to exist and did not suddenly disintegrate; it was so small. And again my mind supplied the answer, "It exists, both now and for ever, because God loves it."

The universe was the temple of her Mother God redeemed by her Mother Jesus.

Four centuries later in the Age of Reason the theism of Julian of Norwich's divine love, the theism of all that is made and sustained, transformed into deism, the deism of all that is designed.

William Paley, archdeacon of the diocese of Carlisle and a strenuous supporter of the abolition of slavery, spent much time in his study, where he tried, as he said, to compensate for his "deficiencies as a churchman." He is celebrated for his *Natural Theology* published in 1802, three years before his death. The subtitle *Evidences of the Existence and Attributes of the Deity Collected from the Appearances of Nature* summarizes the theme of his culminating work.

Notice, wrote Paley, how well-contrived is the eye wherein all parts cooperate to serve a common purpose. Similarly with the hand. Never in all eternity could the eye and hand have arisen by themselves in response to the blind forces of nature. Transparently, they and all other intricate things of the living world were designed by an intelligent deity expressly for the purposes they ably fulfill.

Suppose, said Paley, that while walking on the heath I stumble against a stone. I would not feel provoked into wondering how the stone got there, for it may have lain on the ground for untold ages. "But suppose I found a watch on the ground," a natural conclusion would be ". . . the watch must have a maker; that there must have existed, at some time and at some place or other, an artificer or artificers who formed it for the common purpose which we find it actually to answer, who completely comprehended its construction and designed its use." Around us we see intelligent design, he said, "such as relations to an end, relations to one another, and to a common purpose," and wherever we witness the formation of things, the evidence of God the designer, the clockmaker, stares us in the face.

Paley ably expressed the views of the deists. But nowhere in *Natural Theology* is he particularly original. More than 100 years previously the clergyman-scientist Thomas Burnet had written much the same in his *Theory of the Earth:* "We think him a better Artist that makes a clock that strikes regularly at every hour from the springs and wheels he puts in the work, than he that hath so made his clock that he must put his finger in it every hour to make it strike." Deists in the eighteenth century often used the clock analogy, and the closer they studied the clockwork universe, the more obvious seemed the evidence of intelligent design. As theistic maintenance waned, so deistic design waxed with the advance of science.

A radiolerian (E. Haeckel, *Art Forms in Nature*. Dover Publications, New York, 1974.)

But William Paley's wonderment of all that is designed contributed nothing to explaining how the mechanisms worked.

Deists were left with no recourse other than to peer deeper into the machinery to find out how God had designed it. Inevitably, they and their successors in the nineteenth century discovered that everything was the consequence of natural processes. The stone became no less wondrous than the watch. Even water seemed as purposive in its properties as the eye and hand. If water did not expand on freezing, and therefore ice could not float, the oceans would freeze solid, and life on Earth be impossible. Was this not also evidence of deistic design and foresight?

In the *Bridgewater Treatise*, written by eight distinguished authors and dedicated to demonstrating the "Power, Wisdom and Goodness of God as manifested in the Creation," the chemist William Prout wrote in 1834, "The above anomalous properties of the expansion of water and its consequences have always struck us as presenting the most remarkable instance of design in the whole order of nature—an instance of something done expressly, and almost (could we indeed conceive such a thing of the Deity) at second thought, to accomplish a particular object." The structures and functions of the living world over which Paley marveled were, in fact, potential in the laws of nature, the properties of matter, the elements, and the atoms. The laws and fundamental parts were so designed that they worked naturally to form the wonders of the living world.

Paley saw evidence all around him of an Artificer who fashioned inert matter into elaborate organic structures. But knowledge increased, and the physical universe of Paley's day changed. Design became apparent less in the particular, more in the general, less in the eye and hand, more in the atomic parts. Design was manifest in the differential survival of natural selection, the laws of the heavens, and the fabric of space and time. And this is where we stand today. Most intellectuals in the twentieth century believe the physical universe to be self-regulating, self-determining, and self-sufficing.

☐ ☐ ☐

Why is the physical universe the way it is? Why is it arranged and organized this way and not some other? The answer once frequently given was that perhaps God created many different universes. The other realms of existence were perhaps nothing more than preliminary experiments performed before the final construction of our own. David Hume in his *Dialogue Concerning Natural Religion*, published posthumously in 1779, conjectured that numerous universes ". . . might have been botched and bungled throughout an eternity ere this system was struck out; much labour lost, many fruitless trials made, and a slow but continual improvement carried out during infinite ages in the art of world-making."

Why stop? Might not other universes thereafter have been struck out more splendid than the one we inhabit? Olaf Stapledon in his imaginative book *The Star Maker* describes how the Star Maker devises cosmos after cosmos of increasing sophistication, until each far surpasses our own:

> In vain my fatigued, my tortured attention strained to follow the increasingly subtle creations which, according to my dream, the Star Maker conceived. Cosmos after cosmos issued from his fervent imagination, each one with a distinctive spirit infinitely diversified, each in its fullest attainment more awakened than the last; but each one less comprehensible to me. . . . I strained my fainting intelligence to capture something of the form of the ultimate cosmos. With mingled admiration and protest I haltingly glimpsed the final subtleties of world and flesh and spirit, and of the community of those most diverse and individual beings, awakened to full self-knowledge and mutual insight. But as I strove to hear more inwardly the music of concrete spirits in countless worlds, I caught echoes not merely of joys unspeakable, but of griefs inconsolable.

The ingenuity of Stapledon's psychic world creations matches the ingenuity of his physical world creations. "One inconceivably complex cosmos" made by the Star Maker is of considerable cosmological interest:

> . . . whenever a creature was faced with several possible courses of action, it took them all, thereby creating many distinct temporal dimensions and distinct histories of the cosmos. Since in every evolutionary sequence of the cosmos there were many creatures and each was constantly faced with many possible courses, and the combinations of all their courses were innumerable, an infinity of distinct universes exfoliated from every moment of every temporal sequence in this cosmos.

In this bizarre "exfoliating universe," where freedom of choice is denied by making all possible courses of action mandatory, Stapledon anticipates the many-worlds interpretation of quantum mechanics introduced a few years later by Hugh Everett at Princeton University.

"Time forks perpetually toward innumerable futures," wrote Jorge

Luis Borges, Argentinian essayist and connoisseur of the bizarre, who adopted the exfoliating universe in *The Garden of Forking Paths*.

Pure chance and free choice are unwelcome guests in any rational and orderly scheme. What is not mandatory is forbidden. Imagine you are walking in a wood and come to a place where the path divides into alternate routes. There is no reason to take one path more than the other, and you are free to choose whichever you please. But freedom of choice is an illusion in a rational universe, for all is determined by the inviolable laws of that universe. Liberty to do this or that as you wish, go here or there as you will is intolerable, for it contradicts the rational determinism of the universe you live in. Hence you take both paths.

This is the many-worlds interpretation of the freedom of will. Free choice is an illusion, and all that can happen, must happen.

An atom is a configuration of many waves. When an atom is excited, or disturbed in any way, it evolves as a mixture of waves, each wave representing a possible final state of the atom. The various states vie with one another, and the multiform atomic wave occupies all competing states with different amplitudes. In the analogous situation, the two paths you take in the wood are ghostlike, and you are divided into two ghosts, each following a different path.

The ghostlike world of becoming consists of numerous potential states of numerous futures, and no decision is ever made concerning a particular actual state of a particular future. Everything is potential and becoming in a deterministic phantom world. But when an observation is made, the multiform wave collapses, and a particular actual state comes into being. The observer sees a quantum jump from one state to another. The wavelike world unfolds in multitudinous acts of becoming, and the observed world jumps from one actual state of being to another.

The world of countless atoms forms a vast and intricate probability wave of becoming, full of potentiality, empty of actuality. Parts of the wave collapse into actual states of being whenever we make observations. We cannot predict actualities, only their probabilities, and an element of chance is thus involved. And yet the phantom world of becoming itself is deterministic and free of all caprice.

Whenever an observation is made, some part or other of the wavelike quantum world collapses into actual states. A star is a garden of forking paths that defies the imagination. Each of its wavelike particles follows simultaneously every possible course of action. It is a multiworld of ghostly potentiality having no concrete actuality. Now look at a star through a telescope. The phantom world of the star collapses into an actual state that accounts for the particular photons you observe.

Coming the other way on one of the paths is your friend Mr. Smith. Suddenly your ghostlike two halves collapse into the actual you, and the probabilities of meeting and not meeting Mr. Smith in this case are equal. Perhaps Mr. Smith earlier that day had the choice of either staying at home or going for a walk; one ghostlike half stayed at home, and the other meets one of your ghostlike halves. The probability that you actually meet and observe Mr. Smith in this case is one-quarter.

Trouble now befalls us. An element of pure chance (or probability) is involved in the observed world. We think it not enough that the world of ghostly waves is fully deterministic; in a lucid and rational universe we must have predictable observations, not just predictable probabilities.

In the many-worlds interpretation of free choice and pure chance you actually take both paths in different universes. Mr. Smith actually stays at home in one universe and goes for a walk in another. In one of these universes you meet Mr. Smith, in the other three you do not meet him. The phantom worlds of potentiality are made actual by splitting the universe into many universes each time a decision has to be made. These different universes are not just the phantom worlds occupied by your many ghosts pursuing potential courses of action. You actually take the different paths, each in a real and different universe.

Thus free choice and pure chance are eliminated. You have no choice, because you must take all possible courses of action. In each you look back in time and see the cause of your decision, see the reason why you took a particular path and why Mr. Smith decided as he did.

Hugh Everett in 1957 suggested that the multiform wavelike world of quantum mechanics is the real world, and everything else is mere opinion and illusion. Pure chance and its element of probability are eliminated from the interface of the quantum and observed worlds by making the actual world multiform. The atom of many potential states actually evolves into all states, with each final state in a different universe. From any one universe exfoliates many universes each time a single atom is excited, and because vast numbers of atoms everywhere are constantly excited, each universe branches into vast numbers of universes at every instant. Bryce DeWitt, who has contributed to the many-worlds theory, remarks, "every quantum transition taking place on every star, in every galaxy, in every remote corner of the universe is splitting our local world on earth into myriads of copies of itself. I still recall vividly the shock I experienced on first encountering this multiworld concept."

"To be is to be perceived," said George Berkeley, the bishop. John Wheeler has applied Berkeley's philosophy to the exfoliating universe and suggested that whenever an atom evolves into many potential states, the universe branches into only those universes inhabited by observers. They must contain observers as a precondition for their existence. "It is not that there is a universe and then one has his choice whether he will or will not put man or mind in it," says Wheeler, "it is that in some sense the universe could not exist without mind in it somewhere at some time." Usually, we think of a universe as a precondition for life, not of life as a precondition for a physical universe. The mind undoubtedly is involved, as Wheeler says, but I am inclined to think that it is impossible to place the mind as a distinct entity in the physical universe, or in any universe for that matter.

The meaning and interpretation of quantum mechanics is still far from clear. On the one hand, we have the deterministic quantum world, full of potentiality in multiform states of becoming; on the other hand, we have the observed world of probability, full of actuality in particular states of being. Possibly we have yet to unravel the true nature of time.

□ □ □

Often the why-type question in cosmology is insistent and not easily suppressed. Why is the universe the way it is? One answer, still common today, is because the gods decreed it be this way and not some other.

Why is the physical universe arranged and organized with planets, stars, and galaxies; why is it furnished with certain laws of nature and not others; and why have the fundamental constants (such as the mass and charge of the proton) their observed values? Some of these questions are not scientific in the ordinary sense but cosmological and philosophical. They tend also to be theological, because the answer often given is that the universe, the laws of nature, and the fundamental constants are the product of deistic design.

God designed the universe is one answer. There is another, a sort of Aristotelian answer, revived in recent years and referred to as the anthropic principle.

Lawrence Henderson, a distinguished scientist of broad interests at Harvard University, wrote in 1913 in *The Fitness of the Environment:*

> The fitness of the environment results from characteristics which constitute a series of maxima—unique or nearly unique properties of water, carbonic acid, the compounds of carbon, hydrogen, and oxygen, and the ocean—so numerous, so varied, so nearly complete among all things which are concerned in the problem that together they form certainly the greatest possible fitness.

[247]

Everything is nicely adjusted so that the Solar System and the universe serve as fit and proper places for habitation by life. It is the old design argument of a wonderfully fit universe with the deity left out.

Life exists not because the universe by chance happens to be a fit and proper place for habitation, but the universe necessarily is a fit and proper place because life actually exists. If by mischance the universe were unfit, we would not be here to comment on its unfitness. Our existence places severe constraints on the nature of the universe.

The deistic argument has been turned upside down. We must postulate the existence of human beings, not God, if we wish to understand why the universe is the way it is. This is the essence of the anthropic principle that inverts the deistic design argument.

Arthur Eddington and Paul Dirac showed in the 1930s how the constants of nature and the size of the observable universe have a truly remarkable relation. The constants of nature (such as the speed of light, the electric charges and masses of subatomic particles, the universal constant of gravity, and Planck's constant of quantum mechanics) can be combined in various ways to give pure numbers. These pure numbers are independent of our arbitrarily chosen units of measurement (centimeters or furlongs, grams or hundredweights, and seconds or fortnights) and have the same values as determined by creatures elsewhere in the universe. The values of these pure numbers are universal. Interestingly enough, the numbers obtained in this way divide into two groups; in the first group the numbers cluster around the value of unity, and in the second group the numbers cluster around the enormous value of ten thousand trillion trillion trillion (10 followed by 40 zeros). For example, the ratio of the proton and electron masses is a pure number in the unity group, and the ratio of the electrical and gravitational forces between a proton and an electron is a pure number (roughly equal to 10 followed by 40 zeros) in the second group.

The observable universe has a size of approximately 10 billion light-years; light from galaxies outside the observable universe has not yet reached us. Eddington and Dirac were intrigued by the coincidence that the size of the observable universe is also 10 followed by 40 zeros times as large as a subatomic particle. Thus the constants of nature when combined give pure numbers in the region of 10 followed by 40 zeros, and oddly enough, the size of the universe when measured in subatomic units turns out to be the same large number. The expansion of the universe makes the coincidence surprising; in the distant past the universe was much smaller; in the distant future it will be much larger, and hence the coincidence is a feature of the present era. Is the coincidence fortuitous or indicative of design?

Robert Dicke of Princeton University said it was neither and argued as follows. Stars evolve and manufacture elements heavier than hydrogen. Some stars explode as supernovas and eject into space heavy elements that become embodied in newborn stars and their planetary systems. Life cannot begin until the galaxies have formed and the first stars have evolved and made elements such as carbon that are essential for living creatures. By this time the universe has expanded and reached an observable size that is close to 10 followed by 40 zeros in subatomic units. In the distant future, when the stars have died and living creatures become extinct, the size will be much larger. The existence of living creatures requires that the universe has roughly its present size.

Thus a perplexing coincidence, which at first sight seems to be either fortuitous or indicative of design, turns out to be a natural precondition for the presence of life in the universe. Physicists note the coincidence, said Dicke, and "it is a well known fact that carbon is required to make physicists."

☐   ☐   ☐

The basic properties of the physical universe appear to be arbitrarily determined and fixed in ways not understood. They are the *accidentals* of this world, in Aristotle's terminology. The speed of light, Planck's constant, and other constants of nature such as the strengths of the various interactions (strong, electromagnetic, weak, and gravitational) are not, at present, determined uniquely by any known theory, and they appear to be accidental, even providential. Providential, because if they were different in value, we would not be here discussing the subject.

Let us suppose that many physical universes actually exist, each one complete and self-contained with its own space and time. We may suppose, if we wish, that they all occupy a superspace of some kind; but even so, they remain noninteracting. Among these many universes the accidentals—the unexplained constants of nature—are distributed randomly with various values and arranged in all possible combinations. In some gravity is stronger and in others weaker than in our own; in some the electric charge of the electron (and proton) is larger and in others smaller; and similarly with the rest of the constants. Each cosmos in the multiuniverse ensemble serves as a workshop in which we examine the consequences of the accidentals having values other than in our own. Study of this ensemble of universes leads to a surprising conclusion.

We find that most universes contain only hydrogen. The nuclei of atoms heavier than hydrogen cannot exist (the electric repulsion between protons is too great, or the strong interaction between nucleons is too weak), and these universes lack elements necessary for the formation of planetary systems. They lack, in particular, the carbon, nitrogen, and oxygen necessary for organic molecules and the biochemistry of life. Living creatures, as far as we know, cannot be constructed from hydrogen only, and these hydrogen-only universes are lifeless.

On looking closer at the ensemble we find among the universes capable of having stable heavy elements that many nonetheless are without planets, stars, and galaxies and consist only of a featureless and undifferentiated distribution of hydrogen and helium gas. In these "grin universes," where "all nature wears one universal grin," in the words of Henry Fielding, we find for various reasons (the gravitational constant is too weak, the temperature too high, the expansion too rapid, the cosmic lifespan too short, and so forth) that the prevailing conditions are unfavorable for the formation of astronomical systems.

In many of the universes having astronomical systems the stars are cold and dark. Throughout these inhospitable world systems of perpetual darkness the stellar furnaces remain unlit, and the industry of producing heavy elements from hydrogen and helium stands idle. In a few only the stars shine brightly. In fewer still the stars have luminous lifetimes long enough for biological evolution to occur. (Stars must not evolve too fast and burn out before life originates and evolves from simple to complex states). We notice in these starlit universes where stars evolve over long periods of time that generally there are vast expanses of space. In even fewer we find beneficent sunlike stars emitting radiation of the kind suitable for promoting the biochemistry of life.

Examination of the whole ensemble reveals that life probably exists in only one universe—the one we inhabit—or at most a few practically indistinguishable from our own.

Why is our universe the way it is? The answer is because we are here asking the question. The same answer applies to many probing questions of the why-type. Things are as they are because we exist, and if they were otherwise, we would not be here asking the question. This is the anthropic principle, so-named by Brandon Carter, who has explored the fitness of the cosmic environment as a "reaction against exaggerated subservience to the Copernican principle." The Copernican (or rather Democritean) principle asserts that human beings are not privileged occupants of the universe.

Usually scientists translate existential questions into functional questions. Ontological perplexity is not in their department. Why is the universe the way it is? translates into: How has the universe evolved? But now they have another answer, the anthropic principle, in which an ontological question receives an ontological answer. The principle states: *the universe is the way it is because we exist.* The usual question, Why does life exist? with the usual answer, Because the universe is a fit place, has become the converse question, Why is the universe a fit place? with the converse answer, Because we exist. All other universes of different construction lack the essentials of life, such as long-lived luminous stars and elements heavier than hydrogen, and are therefore lifeless. Only one universe contains living creatures, because the accidentals have come together with the values requisite for life. Our universe is this way and not some other because it contains us who represent and observe it.

It must be admitted that the anthropic principle, with its overtones of anthropocentrism, is in danger of claiming too much. Human beings and extraterrestrial intelligent beings capable of wondering about the design of the universe are not essential to the argument. Wild flowers determine the cosmic design as much as human beings. Butterflies do not exist in universes having other values of the constants of nature. When given a hazelnut we are given all that is made: a universe of galaxies, sunlike stars, vast expanses of space and time, and stable elements made in stars that died before the birth of the Sun. Intelligent beings impose no more constraint on the design of the universe than wild flowers and butterflies. Therefore, the principle is biopic or biometric, rather than anthropic or anthropometric.

□　□　□

Our universe, with its finely tuned properties, may be looked at from two viewpoints. First, from the Aristotelian viewpoint, we see the design parameters as accidental in origin. Because they are truly accidental and randomly determined, this implies that the hypothesized stupendous ensemble of universes actually exists. The constants of nature have in a haphazard cosmogenic manner come together with the precise values requisite for the origin and evolution of life. Countless universes impotent to spawn life are plunged in total darkness or filled with searing light. Of the whole array only one, or a group of similar members, contains living creatures. Here is cosmogenesis on the most profligate scale. We are given, in effect, a law of cosmic selection—the anthropic principle—that discards hordes of universes as unfit to bear life.

Second, from the deistic viewpoint that reigned in the Age of Reason, we see our finely tuned universe as designed by an intelligent deity for habitation by life. The universe contains swarms of galaxies, oceans of space and time, long-lived luminous stars, and finely adjusted constants of nature in order that life shall exist. Granted that the universe is the way it is because life exists, but it is made precisely this way and no other in order that life shall exist. The deistic principle has no need of a wasteland of barren universes, and the ensemble of universes may be discarded as a theoretical fiction that has served the purpose of demonstrating the fitness of the universe we occupy. Why is the universe the way it is? Because God made it that way and no other in order that

life shall exist. Here is the cosmological proof of the existence of God—
the design argument of Paley—updated and refurbished. The fine tuning
of the universe provides prima facie evidence of deistic design.

Take your choice: blind chance that requires multitudes of universes
or design that requires only one.

If indeed our physical universe is but one of many, all configured
differently, several issues clamor for attention. Our physical universe is
self-contained and diversified within its own space and time; others,
assuming that they exist, must also be self-contained and diversified in
their own spaces and times. In no circumstances can we imagine that
these universes are distributed in the space and time of our universe.
Hume and Stapledon are therefore in error when they suppose that numerous universes are created one after another in our particular time.

We have also the dangling question of verification. It is all very well
to adopt a god's eye view while surveying the creation and evolution
of other universes, as in Stapledon's extracosmic theater, but with our
ordinary worm's-eye view, how can we ever verify that any of these
other universes actually exists? Each is self-contained and beyond observation by human beings. The confident assertion in Francis Thomp-
son's *Kingdom of God:* "O world invisible, we view thee, O world intangible, we touch thee, O world unknowable, we know thee," is fine for
the poet but no help to the cosmologist. When postulating other universes we quit the solid ground of observation for the airy heights of
unfalsifiable speculation. As Hume said on this subject, "who can determine where the truth, nay who can conjecture where the probability
lies, amidst a great number of hypotheses that may be proposed, and
a still greater that may be imagined?"

If our universe is a member of a plurality of universes, then which
member is actually ours? Not the Aristotelian, nor the medieval, nor the
Newtonian universe, for each has gone. Nor surely the twentieth-century
physical universe, for in the twenty-first or the thirty-first century, that
may also have gone.

We are inclined to suppose that other universes coexist if only for
the sake of compensating for the deficiencies in our own. Concerned
with the shortcomings of our universe, its disappointments, its lack of
justice, and the mortality of life, we create otherworldly heavens and
utopias from the figments of our world pictures. With the exfoliating
universe we repair the irrationality of the physical world; with the anthropic principle we cover up its inexplicable accidentals.

By supposing that other universes coexist, have we again fallen into the trap of failing to distinguish between universes and the Universe? Might not each of the many universes possess distinctive features that properly belong to the Universe? Any universe, either Aristotelian, Newtonian, or a theoretical model of modern supermarket cosmology, is a product of the human mind and therefore a simulacrum of the Universe.

In no circumstances can we think of the Universe itself as one of many Universes, for the Universe is unknown, inconceivable, and patient of many interpretations.

◻   ◻   ◻

The laws of the universe are the harmonies and conservations found after much observation and thought. They direct our attention and focus our thoughts. We hold fast to the laws until they wear threadbare. They are as impermanent as universes. Where are the laws of 1000 years ago? If this seems too long a span of time, then where can be found a single fundamental law of physics that has survived unscathed for the last 200 years?

Laws of harmony and conservation are generally of two kinds: immanent and imposed. Alfred Whitehead in *The Adventure of Ideas* explains how immanent laws have internal relations and imposed laws external relations.

Laws in the age of magic were mainly of the immanent kind. The world consisted of an activity of indwelling spirits whose thoughts and emotions mirrored those of human beings. Everything was explained in terms of inner states and internal relations.

Laws in the mythic universe were mainly of the imposed kind. Nothing acted of its own accord, and everything was explained in terms of external push-pull relations.

Strange that immanent laws with their subtle internal relations came first. Imposed laws, such as the Commandments and Newtonian gravity, with their simple external relations, came later. Imposed laws must be obeyed, and once understood (or rather remembered), little further thought need be expended in their applications; whereas immanent laws, with their complex internal relations, so flexible and intricate, require the expenditure of much thought in each application. Our everyday ideas are regulated mainly by imposed physical and social laws, perhaps because our cultural heritage descended from the mythic universe in which everything was pushed and pulled by the coercive will of superior social and spiritual beings.

Our modern physical universe has internal and external laws, and much to the concern of the person in the street, its laws are progressively becoming more internal. The once-simple external relations of space and time have given way to the subtle internal relations of special relativity. The law of Newtonian gravity, which Newton thought was theistically imposed, has become immanent in the spacetime deformations of general relativity. Subatomic particles lack simple location, and the laws of the quantum world, immanent in the spirit-dance of atoms, appear to be imposed only when we observe the collective behavior of many atoms.

The ancient atomist universe presents a picture of a physical world fully governed by imposed laws and their external relations. It is a knock-about comedy in which atoms and their assemblies are pushed and pulled by collisional impact. What cannot be explained is dismissed as illusion. This picture is expressed by the equation:

$$\text{universe} = \text{the sum of all atoms}$$

The universe is the sum of all externally interacting discrete entities having simple location in space and time. It is a mechanistic picture having much in common with the mythic universe. It adopts a material world left dead by the retreat of spirits, and the imposed laws of gods have become the imposed laws of nature acting on obedient matter. Utmost simplicity and slavelike behavior of the fundamental constituents are the hallmarks of a world governed by imposed laws. Much of the physical universe is still governed by a calculus of imposed mythic laws and their external relations. Almost instinctively we tear things apart into their simplest components and postulate laws of coercive action when trying to explain what happens.

Inevitably, the laws of the future will not be of the imposed kind with their push-pull external relations. Perhaps they will unify the sub-atomic and cosmic realms with cunningly contrived internal relations at present beyond our comprehension. A doctrine of internal relations opens up the idea of an implicate universe of many aspects, as discussed by David Bohm in *Wholeness and the Implicate Order*, in which each aspect is itself the whole in some special form. In this conjectural universe all things are immanent within one another, and the universe is not a mere sum of bits and pieces ruled by imposed laws. Anaxagoras may have had a glimmering when he declared that "in all things there is a portion of everything."

The notion of internal relations helped to inspire the kalam universe, and Gottfried Leibniz's theory of monads in the early eighteenth century. Monads are the basic components of the universe, said Leibniz, and each monad is itself a complex inner world that reflects the universe. God coordinates these many inner worlds, which are unrelated externally, by means of a system of preestablished harmony.

At about the same time Bishop Berkeley took a different tack and argued that all motion, uniform and accelerated, is relative to the fixed stars of the universe. Ernst Mach, an Austrian philosopher of science, advanced a similar idea in the nineteenth century and proposed that the inertia of a body (its tendency to resist change from a state of uniform motion) is the effect of all the matter scattered throughout the universe. This notion, according to which the universe as a whole supposedly determines local internal properties such as inertia, is now referred to as Mach's principle. The microcosmos in some unfathomable manner reflects the macrocosmos. Recently, the physicist Geoffrey Chew has stated the doctrine of internal relations in its most general form: "Nature is as it is because this is the only possible nature consistent with itself."

The implicate universe may be expressed by the simplistic equation:

$$universe = atom$$

where the equal sign in this case means "is equivalent to." An atom and any assembly of atoms is an aspect or facet of the whole universe and not just an isolated entity. A full understanding of any one aspect implies an understanding of the whole.

Existential questions of the why-type, such as Why are all electrons alike? cannot be answered in a universe of imposed laws and their external relations. The best one gets is in the mythic universe, with the answer, The gods designed electrons that way. Functional questions of the how-type in a universe of imposed laws, dealing with the way things are pushed and pulled into various configurations by fiat and with the way atoms are jostled and juxtaposed into various concatenations by fate are answered much more readily.

In an implicate universe the why-type questions are answered with moderate ease. Why are all electrons alike? Because they all mirror a single aspect of the universe. The whole is a unity of aspects, with each aspect itself the unity. The mechanistic universe tells us how, but not why; the implicate universe, at present only a dream, suggests why, and may one day tell us how.

□　□　□

The anthropic principle, shorn of its many attendant universes, serves as a makeshift scheme of internal relations. It links organisms and the physical universe and serves a useful purpose until better ideas emerge. When given a wild flower we are presented with the design of the universe. Perhaps one day when given an electron we will also be presented with the design of the universe.

But an implicate universe cuts the ground from under the anthropic and deistic principles. Because of our woeful ignorance, everything fundamental seems to us to be either accidental or designed.

The hypothetical implicate universe, which itself is only a universe and not the Universe, has no room for Aristotelian accidentals, because everything interrelates self-consistently. The universe is this way and not some other because it is the only possible universe consistent with itself. The constants of nature are not arbitrary, and we are not free to distribute them with various values over an ensemble of universes.

Nor is the universe designed by an external being in such a way that certain things serve special ends. If we change our strategy and argue that the universe is designed to be self-consistent, then it must be self-complete in every respect, with no need for an external and transcendental deity who imposes design.

My response to all this is that we are neither the precondition nor the consequence of a finely tuned universe. By such arguments we pretend to see image makers in images, and we entrap ourselves in our universes.

We are a part or some aspect of the unknown Universe and are the creators and designers of universes. We, the world makers and world shakers, seek in the cosmic blueprints of our minds to understand all that is made.

16

The Cloud of Unknowing

An unidentified English author of the fourteenth century, who was probably a priest, wrote

> But now thou askest me and sayest: "How shall I think . . . and what is he?" Unto this I cannot answer thee, except to say: "I know not." For thou hast brought me with thy question into that same darkness, and into that same *cloud of unknowing* . . . . For of all other creatures and their works — yea, and of the works of God himself — may a man through grace have fullness of knowing, and well can he think of them; but of God himself can no man think. And therefore I would leave all that thing that I can think, and choose to my love that thing that I cannot think.

Like other contemplative mystics of the Middle Ages the author discovered that thought could not unveil the face of God: "By love may he be gotten and holden; but by thought neither."

Avowed mystics, unlike ordinary contemplative mystics (such as sages and poets), lay claim to intuitional knowledge that transcends ordinary understanding and reject the world of their society. The author of *The Cloud of Unknowing* was not an avowed mystic, not by the standards of the fourteenth century, nor by those of today. Christian, Jewish, and

Muslim contemplative mystics of the Middle Ages ranked among the most advanced thinkers of their time. Thus, for example, Nicholas of Cusa, prince and statesman and sage of the Roman Church, perspicaciously remarked that scientific superstition is the expectation that science answers our every question.

□   □   □

In the West, and wherever else the physical universe now holds sway, sections of the public have caught up with the agnostic intellectuals of the nineteenth century. The educated person has become agnostic. Agnostics believe that the existence of God may be affirmed by faith but not by appeal to reason.

The often-misunderstood word agnostic was first used by Thomas Huxley at a party in London one evening just before the founding of the Metaphysical Society in 1869. A few months later the *Spectator* reported that Huxley "is a great and even severe Agnostic, who goes about exhorting all men to know how little they know." In a subsequent issue of this journal we are told, "Agnostic was the name demanded by Professor Huxley for those who disclaimed atheism, and believed with him in an 'unknown and unknowable' God; or in other words that the ultimate origin of all things must be some cause unknown and unknowable." As a matter of fact, the author of *The Cloud of Unknowing* was an agnostic, like all contemplative mystics.

Whereas atheists deny outright the existence of God, agnostics accept the possibility of God's existence, and with Thomas Huxley deny that God is known and knowable. Agnosticism has two forms: a weak form, as old as philosophy, which holds that belief in the gods stems not from reason but faith; and a strong form, widespread in modern times, which withholds religious faith.

To avert the snare of agnosticism and give comfort to all of faltering faith, the Vatican Council has decreed that "man can know the one true God and Creator with certainty by the natural light of human reason." Natural theology is the branch of cosmology that aims with "the natural light of human reason" to find proofs of the existence of God. Natural theologians assert that God is known and knowable and dispute the claim by agnostics that the ultimate cause is "unknown and unknowable." Natural theologians and agnostics use reason and exclude faith from their discussions.

Few persons in ancient times had the temerity or desire to doubt the reality of gods. Philosophers who sought by disinterested inquiry to determine whether the gods actually existed were generally unpopular and regarded as wasting their time. Natural theology began with the Ionians, as one would expect, and entered the mainstream of Christianity in the Middle Ages. Saint Anselm, archbishop of Canterbury for the last 16 years of his life in the eleventh century, was the first Christian to attempt to prove the existence of God by means of pure reason undiluted with religious faith.

The proofs of God's existence fall into four main groups, referred to as the ontological, moral, cosmological, and teleological arguments.

The ontological argument claims that the existence of God can be demonstrated by means of propositions of self-evident and indisputable truth. The moral argument seeks to demonstrate that this is the best, or least evil, of all possible worlds, and without God there could be no authoritative distinction between good and evil. The cosmological argument endeavors to show that the authentic universe in which we live could not exist without theistic creation and maintenance. The teleological or design argument aims to prove that the authentic universe fulfills deistic intention; some aspects of this last argument were considered in the previous chapter.

A miscellany of proofs, devised by Thomas Aquinas and referred to as the Five Ways, illustrates the thrust of natural theology:

> *Things are in motion, hence there is a first mover.*
> *Things are caused, hence there is a first cause.*
> *Things exist, hence there is a creator.*
> *Perfect goodness exists, hence it has a source.*
> *Things are designed, hence they serve a purpose.*

At the end of each proof Aquinas said, "all understand that this is God," or words to that effect. The first three belong to the cosmological argument, the fourth belongs to the moral argument, and the fifth to the teleological argument. Aquinas had little patience with the ontological argument and thought it logically unsound.

The ontological argument was promoted by Anselm, who hit on the idea of defining God as "that being than which nothing greater can be conceived." He argued that the reality of God is obviously greater than the idea of God, and according to his definition, God is therefore undoubtedly real and must exist. Mortimer Adler in his popular book *How to Think about God* tries to make the point clear: "to think about the supreme being as merely an object of thought, is not to think of the supreme being at all, but of a lesser being," an imaginary being who falls short of Anselm's definition.

Anselm was delighted with his proof. It misfires, however, because the conclusion is implicit in assumptions not covered by his definition. Almost two centuries later, Thomas Aquinas rejected Anselm's ontological argument and said, granted that the supreme being can be so defined, "it does not follow that what the name signifies actually exists, but only that it exists mentally."

We might, parodying Anselm, define Mephistopheles as that being than which nothing greater in evilness can be conceived. Although the existence of the Devil would undoubtedly be a greater evil than the mere idea of the Devil, this definition fortunately does not establish the reality of such an unwelcome being. Or we might define the physical universe as that thing than which nothing greater in size can be conceived, but again this would not demonstrate that the universe is necessarily of infinite extent, when it may actually be finite and spatially closed.

René Descartes also proposed an ontological proof of the existence of God. He wrote in a letter, "I dare to boast that I have found a proof of the existence of God which I find fully satisfactory and by which I know that God exists more certainly than I know the truth of any geometrical proposition." His argument, stated briefly, asserts that God, who is perfect, must exist because existence is an indispensable element of perfection. A perfect supreme being cannot be merely an imaginary being.

In his *Third Meditation* Descartes wrote, "I shall now close my eyes, stop my ears, turn away all my senses, even efface from my thoughts all images of corporeal things, or at least, because this can hardly be done, I shall consider them as being vain and false." By introspection he found, "I am a thing which thinks, that is to say, which doubts, affirms, denies, knows a few things, is ignorant of many, which loves, hates, wills, does not will, which also imagines, and which perceives." By self-contemplation Descartes had already concluded that his own existence was beyond all possible doubt: "I think, therefore I am." In the *Third Meditation,* he continued,

> There remains then only the idea of God, in which I must consider whether there is anything which could not have come from me. By the name of God I understand an infinite substance, eternal, immutable, independent, omniscient, omnipotent, and by which I and all other things which exist (if it be true that any such exist) have been created and produced. But these attributes are so great and eminent, that the more attentively I consider them, the less I am persuaded that the idea I have of them can originate in me alone. And conse-

quently I must necessarily conclude from all I have said hitherto, that God exists. . . .

He perceived himself as an imperfect being who nonetheless had notions concerning the nature of perfection. Whence came these notions of perfection? Not from himself, nor any other imperfect being; therefore, they must have come directly from God, and nowhere else. Starting with the singular fact that he existed, Descartes came to the conclusion that God's existence was equally beyond doubt.

Descartes then cast his introspective net more widely and found that he himself possessed a soul as a result of God's existence. Other human beings, he conceded, must also have souls. But not ordinary creatures, because "there is nothing which leads feeble minds more readily astray from the straight path of virtue than to imagine that the soul of animals is of the same nature as ours . . . and after this life we would have nothing to fear or hope for any more than the flies and ants." The creatures of the animal kingdom were soulless machines placed on Earth for the convenience of mankind. It seems that when faith intrudes, staining the purity of natural theology, the conclusions drawn in the name of reason tend to be limitless.

Immanual Kant rejected the Cartesian argument on essentially the same grounds used by Aquinas against Anselm: "The concept of a supreme being is in many respects a very useful idea, but just because it is a mere idea, it is altogether incapable by itself alone of enlargening our knowledge in regard to what exists. It is not even competent to enlighten us as to the *possibility* of any existence beyond that which is known in and through experience." The force of the ontological argument resides in its appeal to religious belief, and as an intellectual exercise in pure reason there seems little doubt that the conclusions reached are generally unwarranted. Agnostics remain unconvinced, and the argument therefore fails in its main purpose.

☐   ☐   ☐

The moral argument, which was especially favored by Kant, takes for granted the premise that ethical principles and moral standards are the province of religion and leads to the conclusion that God is the source of all distinction between right and wrong, between good and evil, that without God there can be no perfect goodness. The argument is not accepted by agnostics, who regard the social evolution of moral codes as more natural.

The scorpion's sting that paralyzes the argument is the abundance of evil in the world to which religious institutions and their ardent devotees have made disproportionate contributions throughout history. Embarrassing biblical statements of the kind, "I am the Lord . . . I make peace and create evil," make it difficult if not impossible to construct a consistent and convincing moral argument on the basis of traditional religious teachings.

David Hume in his essay *The Immortality of the Soul* poured scorn on the argument: "Let us now consider the moral arguments, chiefly those derived from the justice of God, who is supposed to be interested in the future punishment of the vicious and the reward of the virtuous. But these arguments are grounded on the supposition that God has attributes beyond what he has exercised in the universe with which we are acquainted." A century later, in *The Utility of Religion*, John Stuart Mill argued that in no way "can the government of nature be made to resemble the work of a being at once good and omnipotent." If the supreme being is accountable for all this wretchedness, then that being is either not perfect goodness or not all-powerful. Bertrand Russell in *Why I Am Not a Christian* rams the point home when he says, we "could take the line that some of the gnostics took — a line which I often thought was a very plausible one — that as a matter of fact this world that we know was made by the devil at a moment when God was not looking."

Plausibly, the moral codes by which we live and the ethical sparks that illumine our world pictures are as old as *Homo sapiens*. In the tens and hundreds of millennia of the magic and magicomythic ages the social groups that consisted of criminally disregarding individuals had little chance of surviving. The ancient codes in the mythic universe became ordinances decreed by almighty gods. The gods legislated the laws and acted as lords of justice, and those who defied the gods by committing sin deserved and received the severest punishment.

And now in modern society what have we? The gods have fled from the physical universe, taking with them the seals of authority affixed to our moral standards. The ancient codes that distinguished between right and wrong, between good and evil, are now mythical, the objects of derision, often irrelevent in political and legal deliberations. The remission of sin has become the denial of sin. We are taught that no one is evil; the cause is hereditary and environmental or the result of mental illness; the fault is never our own. All is not lost, however, for the codes of behavior of primitive men and women linger on, and deep within ourselves we know what is right and wrong, what is good and evil. The codes are still active, preserving society, and they remain our primary defense in the survival game.

The moral proof of the existence of God, once the most persuasive of all arguments, has become the least convincing in an age that takes for granted that the physical universe is without intrinsic ethical content.

The cosmological argument establishes proof by showing that the authentic universe is neither self-creating, self-sustaining, nor self-sufficing and demonstrates that the existence of God repairs one or more of these deficiencies. The closer we identify ourselves with the beliefs of the mythic universe, the more compelling becomes the cosmological argument. The argument nowadays relates to matters of a scientific kind, in which natural theologians are generally not very knowledgeable. The containment of physical space and time in the universe, for example, has greatly altered the nature of cosmogenesis.

In this book all variants of the cosmological proof are off-limits and receive short shrift. I hold that it is impossible to find proof of the existence of God within the framework of a particular universe, for all universes are the handiwork of human beings.

John Laird in his book *Theism and Cosmology* takes the first step in this direction when he remarks, "One of the principal obstacles that beset all arguments from the world to God is the doubtful legitimacy of arguing from the relations or connections *within* the cosmos to a similar relation or connection *between* the cosmos and some transcendental being." The obstacle is actually worse than Laird states. Like most theologians, he speaks of relations within *the* cosmos rather than within *a* cosmos. We may, if we wish, relate God and the Universe, both of which are unknown, but it is unfitting and even illogical to relate God and a universe. The subject of God and the Universe, and gods and universes, is discussed shortly, when we consider some further aspects of the ontological argument.

The teleological argument aims to demonstrate that the universe is deistically designed to serve special purposes. Immanuel Kant, who took great interest in the proofs of God's existence, said the argument from design is "the oldest, the clearest, and the most in conformity with the common reason of humanity." Many scientists, when they admit their views, incline toward the teleological or design argument. The design argument now competes with the anthropic principle.

The teleological argument fails because it is unfitting and illogical to hold God responsible for that which human beings are the sole authors. The beautifully designed medieval universe, as we now know, was undeniably of human and not deistic design.

□   □   □

Anselm's definition of God as "that being than which nothing greater can be conceived" is interesting, for its bestirs the following train of thought. (I have since discovered that similar ideas were expressed in William Hamilton's article, "Philosophy of the Unconditioned," published in *The Edinburgh Review*, 1820.)

We are given two mutually exclusive realms: the first contains the set of all things at present conceivable, whether real or fanciful, and the second contains the set of all things at present inconceivable and beyond the reach of human comprehension.

General relativity, quantum mechanics, quarks, little green men, and the cow that jumped over the moon are members of the first set. We cannot, of course, give specific examples of members of the second set, but from our knowledge of the history of science and our realization that we still do not know and comprehend everything, it would be rash for us to deny the possibility of the second set. The inconceivable in the course of time becomes conceivable; thus general relativity and quantum mechanics were inconceivable in Anselm's day, and much that is now inconceivable will become conceivable in the universes of the future. The familiar first set consists of all conceivable things, including the concept of inconceivability, whereas the unfamiliar second set consists of all inconceivable things, including the cloud of unknowing.

In which of these two realms do we place God? If in the first, amidst the conceivable, we must agree with Anselm that God exists at the limit of what can be conceived in greatness. In that case reason alone is unable to assure us that greater things than God are not members of the second realm. But if in the second realm, amidst the inconceivable, in a cloud of unknowing, we are denied any means of measuring relative greatness.

It seems that the battle is lost and we are now in flight, pursued by saucy agnostics armed with Anselm's double-edged definition, who demand to know whether God occupies the first or the second realm. As a last-ditch defense we are forced to modify Anselm's definition into something like, "God is all and inconceivable." God is all, necessarily, for other things of an inconceivable nature might otherwise be greater.

Anselm's definition has at least the merit of making us realize that we are considering what can and cannot be conceived by the human mind. Though we grant that the existence of a supreme being is conceivable, the concept by itself fails to place the supreme being in the first realm, because the actual nature of that being is inconceivable and therefore belongs to the second realm. God unavoidably is diminished when adorned with conceptual detail. As Aquinas said, the divinity "exceeds by its immensity every form which our intellect attains."

Owing to human limitations, we are compelled in our role as natural theologians to place God in the second realm, amidst the inconceivable, and we are then left with no more than the name of an ungraspable entity of problematical existence. The familiar generalizations associated with the name are at best stock-in-trade metaphors and doubtless inferior to others yet to be conceived.

Anselm's terminology tempts us to define the Universe as that thing than which nothing greater can be conceived. This definition, however, is not good enough. It is insufficient in scope, because the Universe lies in the second realm and its nature exceeds the limits of what can be conceived. The Universe is partly known by means of the present universe and may be better known in the universes of the future. The only acceptable definition states that the Universe is all-inclusive in scope and inconceivable in nature.

Behind our last-ditch defense we witness an alarming sight. On to the field of battle marches cosmology, whom we had thought to be an ally, who bluntly declares that we are trespassing on territory that by ancient and inalienable rights belongs to cosmology: the Universe itself and not God is "all-inclusive and inconceivable."

We are faced with two candidates competing for the same definition: God is all and inconceivable, and so is the Universe. Guided by pure reason and nothing else, we are unable to contrast and compare God and the Universe. Both are off-limits and beyond reach of comprehension.

It is the custom, inherited from the mythic universe, to think and speak of God as a male being — as Him — exalted in the highest degree and endowed with superlative human characteristics, which is where faith comes in the door and reason goes out the window. For the sake of a personal and satisfying relationship of the kind human beings once had with the spirits and godlings of long ago, God is personified and brought into the realm of the conceivable. It is then relatively easy, as in the medieval and other monotheistic versions of the mythic universe, to contrast and compare a personal God with other conceivable things.

All apparent differences between God and the Universe arise as the result of diminishing the status of one or the other by bringing it into the first realm. When both are brought into the first realm (the realm of the conceivable) we profess that one is personal and the other impersonal; when both are left in the second realm (the realm of the inconceivable), where they rightly belong, such a distinction is quite impossible. Pure reason by itself tells us that we must leave God in the second realm without identifying features, anthropomorphic or otherwise, possessing nothing more than the same definition as that of the Universe.

□   □   □

The myriad gods are models of God in much the same way as the myriad universes are models of the Universe. The universes are the masks of the Universe, and the gods are "the masks of God" (to borrow a phrase used by Joseph Campbell). Even a personal and loving supreme being is a mask, a god, conceived and figured by the human mind. The gods and universes are grand unifying ideas, great schemes of organizing thought, occupying the realm of the conceivable. When God and Universe both occupy the realm of the inconceivable, we know nothing more than our ignorance.

We may arrange our thoughts in such a way that a particular god is greater than a particular universe and is the creator *ex nihilo* of that universe. But as cosmologists trying to delve more deeply, we are not at liberty to do the same with God and Universe, because both are unknown and wholly inconceivable. Furthermore, it would be absurd to adopt the halfway argument that God is the creator of a particular universe — the one happening to be in vogue at the moment — because universes are creations of the human mind. It would be equally absurd to argue that a particular god — the one happening to be in vogue at the moment — is the creator of the Universe, because gods also are creations of the human mind.

The difficulty with most arguments in natural theology is that God and gods, and Universe and universes, are never properly distinguished. Thus a not uncommon argument is that "the cosmos does not have in itself a sufficient reason for its own existence." It is then shown that the existence of a particular god remedies that deficiency. But how can we, as earnest investigators, hope to penetrate to the heart of the subject by discussing theistic necessity in this way, when the masks of God and the masks of the Universe forever change and are mere representations of the unknown and unknowable?

It is always within the power of human wit to show that a god is greater than a cosmos, that the existence of the latter depends on the former, and hence the former must exist. When we realize that it is the existence of the Universe that must be explained, such arguments fail, because the Universe is inconceivable and its existence may require no more justification than the existence of God. Those who call on their gods to explain how universes originate must by symmetry expect others to call on their universes to explain how gods originate.

The trouble with theoretical religion in general is that it fails to distinguish between a model of a thing and the thing itself; the gods are models of God and the universes models of the Universe. When God is related to a specific universe, as in monotheistic religions, either as a transcedent or immanent being, the notion of God is diminished and then in peril of being discarded with the advent of a mechanistic self-running universe.

We may, if we wish, equate gods and universes (always stressing that both are models), or equate God and the Universe (always stressing that both are unknown and inconceivable), but we cannot equate a god and the Universe, or equate a universe and God.

We are unable to enter the realm of the inconceivable and there hold discourse on things beyond comprehension. Pretense of this sort consists of bandying around words lacking substantive meaning and tossing about notions of relative magnitude lacking a calculus of measurements. The terminology of the eternal, infinite, omnipotent, and omniscient, though impressive, is meaningless unless, of course, we invent models and thus return to the realm of the conceivable, with its charade of universes and gods.

God and the Universe are defined as all-inclusive and inconceivable, and we are faced with the problem of a redundancy of all-inclusive things of an inconceivable nature. Our definition contains nothing to prevent us from speculating that both are plausibly one and the same thing. Such a startling hypothesis, the trimmed result of Ockham's razor, has the merit of conceptual economy. Moreover, it has the advantage of unifying things beyond the limits of human understanding that possess rival claims to universality. The Heraclitean idea of Mind and the whole of Nature being one and the same was advocated by the Dutch philosopher Baruch Spinoza in the seventeenth century. Johann Goethe, poet and sage, was keen on the same thought:

> Nature! We are surrounded and embraced by her: powerless to separate ourselves from her, and powerless to penetrate beyond her. . . . She has always thought and always thinks; though not as a man, but as Nature. She broods over an all-comprehending idea, which no searching can find out. . . . She has neither language nor discourse; but she creates tongues and hearts, by which she feels and speaks. . . . She is all things.

Many years later, at the age of 69, Goethe said of his earlier views, "it may pass as a jest, with a bitter truth in it."

Pantheism is the belief that gods in the form of great nature spirits are immanent within but not transcendent over the universe. A pantheistic world view lies between the magic and mythic universes, and where it lies on the keyboard, either toward the magic end or the mythic end, depends on the extent of the unifying power ascribed to the immanent spirit gods. A pantheistic world view may be polytheistic or monotheistic.

[267]

Let me hasten to add that Spinozistic philosophy is not just another fusion of models, not mere pantheism, as so often said, for it equates inconceivable things of all-encompassing scope. God and Universe are not immanent in each other, as in pantheism, for both are identical and without distinction. "God is all" takes on the wider meaning of "all is God." Unifying God and the Universe yields a *UniGod*. "All is the Universe" becomes "The Universe is all." The ultimate reality we seek to understand by means of our universes is not a mechanistic world of dead matter that excludes the conceiving mind and all that we associate with the names of gods.

Theological and cosmological traditions in Western society prohibit equating gods and universes. In our theologies the favored gods are confused with God; in our cosmologies the favored universes are confused with Universe, and our concepts of gods, on the one hand, and of universes, on the other, have diverged to the stage where they apply to totally dissimilar things. We dare not tamper with them, and we shudder at the thought of equating models of God to models of the Universe. Because of his philosophical views, Spinoza became a religious and intellectual outcast, excommunicated by his Jewish community that had sought refuge in Holland from the Spanish Inquisition, and shunned by philosophers even to this day.

Rejection of the possibility of a God-Universe or UniGod perhaps explains why we find ourselves in desperate need of proofs of God's existence. Long ago human beings abstracted from the natural world all that they ascribed to the gods, leaving the world dead; now the gods have fled into a surrealistic world of improbable existence, taking with them the half of the natural world that we call divine. We ourselves have transformed God into a fiction that cannot be proved true.

Who doubts the existence of the unknown and unknowable Universe of which we are a part? The history of cosmology reveals many universes, and when we extrapolate from the past to the future, we think it not unreasonable to suppose that numerous universes will exist in the tens, hundreds, and thousands of millennia to come. And we must not forget the universes of extraterrestrial intelligent beings. Each universe masks the Universe whose reality is beyond all doubt.

Given that the Universe and God are one and the same unknown and unknowable reality — the cloud of unknowing — then we cannot doubt the existence of God, because the existence of the Universe is beyond doubt. This, I am inclined to think, must be the ultimate ontological proof of the existence of God. When pressed to its limits, the ontological argument leads to no other conclusion. If we recognize that God and Universe are interchangeable names referring to the all-inclusive and inconceivable, then the reality of God is beyond doubt. Oddly enough, this proof of the existence of God springs from agnostic soil.

□ □ □

Cosmology and theology are linked by six cardinal dualities. The first two comprise compatible things and are what might be called the commensurate dualities. They are: God and Universe, and gods and universes. The other four comprise incompatible things and are what might be called the incommensurate dualities. They are: God and gods, Universe and universes, God and universes, and Universe and gods.

The first of the commensurate dualities contrasts God and Universe. The existence of God and Universe is conceivable, but otherwise both are unknown and unknowable. According to the Spinozistic hypothesis, they are plausibly one and the same, and cosmology and theology thereby recover an original partnership in an enterprise that seeks to unmask the ultimate. The hypothesis has the merit of theoretical economy, and as Newton said, "Nature is pleased with simplicity, and affects not the pomp of superfluous causes."

The second commensurate duality contrasts the gods and universes of a myriad-godded cosmic vortex. On the one hand, we see the universes, wondrous and mind-gripping; on the other, we see the gods, marvelous and inspiring, spinning cosmic fabrics and controlling the gigantic machinery of nature. The universes are masks of the Universe, and the gods are masks of God. Both sets of masks are compatible insofar as they comprise models lying within the realm of the conceivable. Pantheism weaves the two together into a unigod. All is well, provided we resist the strong temptation to treat any one mask as the true face. Those who flout this basic rule of cosmology and claim to know what the Universe is because God has told them, are prisoners in their models.

The first of the incommensurate dualities — God and gods — is as old as the mythic universe in which polytheistic and monotheistic masks were venerated as the true face. Misidentification of the theistic model with the thing itself is as rife today in the world's vast populations as at any time in the past. It is undoubtedly the source of confused thinking in many of the arguments that purport to prove the existence of God.

The second incommensurate duality — Universe and universes — is as old as cosmology and is the main theme of this book. Wherever we alight in the history of cosmology, we find the contemporary universe mistaken for the Universe. This misidentification also is as rife today as at any time in the past.

The third incommensurate duality — God and universes — is the source of confused thinking that occurs in the cosmological and teleological arguments. The origin of this duality is not difficult to understand. Spirits in the age of magic retreated and became godlings, then gods, and finally the unified all-powerful godhead of the mythic universe. Spirit forces were withdrawn from the world and lodged in the realm of God. On the one side, we have universes conceivable and inglorious, and on the other, we have God inconceivable and glorious. Natural theology tends to accept these incommensurate propositions and uses them as a springboard for demonstrating the logical necessity of a supreme being. A feeling of urgency, tinged with despair, lies behind the demonstrations. The glorious has been extracted from the world and given to a supreme being, and without the existence of that being we are left with precious little, other than the struggle of science seeking to rediscover the glorious, which for many persons unfortunately amounts to virtually nothing, because they misunderstand and even dislike science.

The fourth and last of the incommensurate dualities — Universe and gods — is the source of confusion in the material mechanistic universe. Materialists recognize that the gods are elaborate anthropomorphisms and regard them as a hangover from the mythical intoxications of the past. They commit, however, the error of mistaking their wonderful and rich mechanistic universe for the Universe, and the gods are then naturally discarded as being incompatible with the Universe. But the gods, which are models of God, stand on the same conceptual level as the mechanistic universe, which is a model of the Universe.

□   □   □

We feel an urge to believe in God, and this desire derives not from arguments of pure reason. Articulated in holy lexicology, hallowed by tradition, the urge is admittedly irrational within the context of the physical universe. Rationalists resist the urge on the grounds that it emerges from the jungle of misguiding elements in our cultural heritage. Many rationalists and most agnostics realize that absence of proof is not proof of absence, and therefore they take an occasional interest in arguments claiming to show that we have the necessary and sufficient reasons for believing in the existence of a supreme being.

One may legitimately argue that as a result of rejecting the notion of gods, our views concerning reality have become pallid and inane, and our views concerning life itself devoid of satisfying social and personal meaning. Our cultural heritage impels us to believe in God or something similarly all-inclusive and inconceivable, for it has stolen from the phenomenal world the very elements essential for a life of significance and given these elements to the gods who have the function of sharing them with us.

One has only to imagine the home with its own house god or hearth goddess, who emanates an ambience of warmth and friendliness, wards off danger, and is acknowledged by libations and flowers, for the home to seem more secure and restful. This fanciful pagan illustration shows how deep within us is a strong urge to live in fellowship with the gods, and when we deny their existence and live in a godless universe, we are left with a residue of unfulfilled yearning.

In *The First Three Minutes* Steven Weinberg makes the remark, "The more the universe seems comprehensible, the more it seems pointless." The overthrow of the medieval universe, with its unison of all experience, followed by the rise of the mechanistic universe, has left the minds of men and women flittering aimlessly like moths among the mechanisms.

Belief in God or something similarly all-inclusive and inconceivable enables us to cling to a sane sense of proportion and view with equanimity the prospect that our universes are not the Universe and never can be. Belief in an unknown and unknowable God or Universe or UniGod at least counsels a sense of humility and reminds us constantly of the mystery of the inconceivable. In this way the gods within us will be placated.

If we can think that all is far from known, and God is perhaps the Universe, then without further intellectual commitment we avoid the dreariness of atheism. By equating God and the Universe we give back to the world what long ago was taken away. The world we live in with our thoughts, passions, delights, and whatever stirs the mortal frame must surely take on a deeper meaning. Songs are more than longitudinal sound vibrations, sunsets more than transverse electromagnetic oscillations, inspirations more than the discharge of neurons, all touched with a mystery that deepens the more we contemplate and seek to understand.

Alfred Whitehead once wrote, "the theme of cosmology is the basis of religion." The converse statement, the theme of religion is the basis of cosmology, rings with equal truth. This perhaps explains why interest in cosmology has grown in the twentieth century while commitment to religion has declined. Both have the same basic theme, with the one complementing the other. "Science without religion is lame, religion without science is blind," said Albert Einstein, and from his occasional remarks one gathers that he favored a Spinozistic form of religion.

☐   ☐   ☐

On occasion during my school days Canon Morrow sauntered across the cathedral close and took over a scripture lesson in one of the classrooms. In front of the class, in a ringing Church-of-England voice and with dramatic gestures, he brought the past to life. Acting the part of a Roman centurion, he would leap aside from Queen Boadicea's chariot as it swept by on the field of battle, and with arm upraised hurl his imaginary javelin at a barbarian chieftain, and then, skipping across the room to act the part of the chieftain, turn around and stagger forward as the javelin pierced his breast. I sat riveted in my seat as one of the enrapt pupils. After the canon had left to pursue his other duties, our teacher would resume control, sometimes making guarded comments that have since led me to realize he was an agnostic.

By modern standards education at a cathedral school many years ago in England was totally mistaken and grossly misdirected. But it gave me dragons to slay and set me on paths of inquiry that have lasted a lifetime. I thank the gods I escaped the inanities of modern education.

Do not deny the gods. Fight the gods if you will. But grovel, and in their contempt they will scoop you into the holy mincing machine of incarnadine wars. Hate them! Curse them! Though they may crush you, they will not despise you. But if you ignore them, then beware! For in their anger they will inflict upon you nameless horrors of body and mind. Only fools deny the hereditary gods that live within us.

17

Learned Ignorance

In his work *On Learned Ignorance,* written in 1440, Nicholas of Cusa argued that though the darkness of unlearned ignorance disperses in the light of knowledge, there is another side to ignorance, which he called learned ignorance, that grows with knowledge and wisdom. "No man, even the most learned in his discipline, can progress farther along the road to perfection than the point where he is found most knowing in the very ignorance that characterizes him; and he will be the more learned, the more he comes to know himself for ignorant." Consider the unlearned, unaware of their ignorance, who think they know everything! As knowledge increases, ignorance decreases, but this kind of ignorance — unlearned ignorance — is no more than lack of knowledge. With knowledge comes an awareness of ignorance — learned ignorance — and the more we know, the more aware we become of what we do not know.

Cardinal Nicholas of Cusa's principle of learned ignorance comes as a surprise. "Knowledge is strength," says the proverb. We seek learning at the feet of teachers and endeavor to widen our horizon of knowledge. Education dispels ignorance, we believe, and learning ennobles the mind. Perplexing issues that arise in the learning process receive attention within the corpus of growing knowledge, and whatever continues to puzzle is either irrelevant or deferred to a more advanced stage of learning. Then later, as learned scholars, some of us take to research, groping ahead, seeking to advance the frontier of knowledge.

What is wrong with education and research, we exclaim; what is the cardinal lamenting about? Let me try to explain in modern language how I see the principle of learned ignorance.

As the horizon of a person's knowledge widens, or the frontier of a society's knowledge advances, new facts and ideas cast shadows of doubt over old facts and ideas, and the whole domain of previously acquired knowledge must be continually revised and reinterpreted. Gnawing doubts become constant companions. Learned ignorance is an awareness of one's ignorance, and like entropy, it seems never to decrease. It urges us to seek certainty in greater knowledge, which when attained unfailingly creates further uncertainty. The more we know, the more aware we become of what we do not know.

Tentatively I offer the suggestion that the first law of knowledge is the conservation of ignorance. (The first law of thermodynamics is the conservation of energy.) As unlearned ignorance decreases, learned ignorance increases.

A society at first may not be heavily burdened with an awareness of ignorance. With confidence its members seek to remedy the inadequacies of the known. Each new fact and idea adds to the montage of knowledge. Soon the end of knowledge seems in sight. Always, however, some facts and ideas cast deep shadows of doubt, thereby quickening the search for more reliable knowledge. But newly acquired knowledge eventually reaps the harvest of further doubt. The universe is never perfect and free of contradiction, never complete and in full accord with observation. The greater our knowledge, the more apparent becomes the imperfection and the incompleteness of the universe we seek to understand. As the light of knowledge spreads and brightens, the shadows of learned ignorance gather and darken.

Protracted struggles develop between various sections of society, between traditionalists and noncomformists, between unlearned ignorant with authority and learned ignorant without authority. Eventually, the overburdened edifice of traditional knowledge collapses in ruins, often amidst the foundations of a rising edifice of new knowledge.

I hesitantly suggest for the second law of knowledge that learned ignorance tends to increase faster than knowledge. (The corresponding second law of thermodynamics states that whenever possible entropy in an isolated system tends to increase.) What evidence is there for this strange and bewildering law that says awareness of ignorance grows at least as fast as knowledge? My answer is the following.

Universes evolve and generally become more complex and organized. Learned ignorance, which is one of the main agents causing universes to evolve, must increase at least as fast as knowledge. For if it increased slower, then in the course of time what is known to be unexplained would diminish in proportion to what is believed to be explained, and as a result societies would steadily become more secure and their universes more durable. History shows that usually the opposite happens: primitive universes (for example, magicomythic) endure for long periods of time, whereas complex universes (for example, Aristotelian and Newtonian) endure for relatively short periods. It seems reasonable to say that learned ignorance grows faster than knowledge, or at least as fast, otherwise universes would progressively become more secure and enduring. The advance of knowledge would slow down and finally come to a halt. But history shows that the pace quickens, not slackens.

The penalty of knowledge is doubt. The greater the knowledge, the greater the doubt, and hence the greater the urge of many members of a society to banish doubt by the attainment of more reliable knowledge. The huge rock of Sisyphus, which he must forever push up the mountain, and which persistently rolls back before he succeeds in getting it to the top, like learned ignorance, grows larger and larger.

The people in the future, with their vast knowledge, will probably bear burdens of learned ignorance immensely heavier than our own.

□   □   □

Why pursue knowledge and its burden of learned ignorance, why not remain happily unlearned, untroubled by gnawing doubts and uncertain thoughts? An individual is always free to make this decision, but never a society. Individuals are free to withdraw into the security of their irrational world pictures; but a society, to survive and function effectively, is constrained by the rationality of its universe. Individuals may drop out; a society must carry on.

Learned ignorance presents a dilemma. Realization that knowledge leads to doubt, and doubt in turn spurs the search for more knowledge, leading to further doubt, creates frustration and despair. It tempts individuals to renounce the Sisyphean struggle for knowledge, to rest happily in a state of unlearned ignorance, and to take comfort in avowed mysticism. Here is one horn of the dilemma: reject the endless quest for knowledge, become mystics, and substitute for the rational universe of society the unfalsifiable mythic elements of vestigial universes.

But mysticism, however reassuring, belongs to personal world pictures and not to the impersonal universe of society. The universe, as a common denominator, must always seem rational in accordance with its principles and terms of reference. Mysticism is credible and tolerable within world pictures, maybe even necessary in moderation, but always incredible and intolerable when foisted on the universe. Everybody's mysticism differs, and a universe at the mercy of multiple mysticisms becomes a doomed Tower of Babel.

Alternatively, individuals escape frustration and despair by having faith in the prevailing universe of their society. They think that the universe is obviously the Universe or a very good approximation, and they gain confidence from the belief that the journey's end is reached or in sight. At last, the heroic Sisyphean struggle is at an end; the rock has reached the top of the mountain. Here is the other horn of the dilemma: believe that the end of knowledge is reached or in sight, and find encouragement in the conviction that the known universe is the Universe.

We flee from our individual burdens of learned ignorance into avowed mysticism, rejecting the universe of society, or into unquestioned faith in the universe, believing it to be the Universe. Dogmatism triumphs in either case. Unlearned ignorance, when adopted by choice as an attitude of mind, fosters mysticisms, dividing and destroying a society when practiced widely, or fosters orthodoxies, causing social revolution when practiced inflexibly.

One resolution of the dilemma is skepticism: the denial of knowledge and the rejection of the cloud of unknowing. Beyond the dissonant tumult lies nothing — nothing worth seeking, nothing worth knowing — everything is ultimately meaningless and pointless.

My resolution is to regard each universe — or mask of the Universe — as an aspect of reality. All universes are real, and the Universe is patient of many interpretations. Undenied learned ignorance leads to wonderment in a quest of endless wonderment.

Maybe there is a third law of knowledge: unawareness of ignorance, which is unlearned ignorance, can never be zero. (The corresponding third law of thermodynamics states that the temperature of a system can never be reduced to zero.) One cannot question everything. The burden of learned ignorance is supportable when we derive from unlearned ignorance unquestioned faith in the existence of the cloud of unknowing, whether it be God, the Universe, or UniGod.

In 1910, 1912, and 1913 Bertrand Russell and Alfred North Whitehead published the three volumes of *Principia Mathematica*. In this monumental treatise they aimed to construct a rigorous system of mathematical logic that would be complete and consistent (free of contradiction), such that all mathematical statements expressed in the context of the system could be demonstrated to be either true or false by means of its symbolic apparatus. Their aim was admirable and seemingly logical.

But in 1931 Kurt Gödel, a young Austrian mathematician, published a short paper, "Formally undecidable propositions of *Principia Mathematica* and related systems," and dashed all hopes of ever realizing a complete and consistent mathematical system. Any system claimed to be complete contains inconsistent elements.

A mathematical system is constructed by carefully devising a set of interlocking postulates. With the deductive logic established by these postulates or axioms the mathematician then determines whether various propositions are either true or false. Many persons, for instance, are vaguely aware of a set of ideas sufficiently complex and complete to determine the truth of the propositions $1 + 0 = 1$ and $1 \times 0 = 0$, but not $1 \div 0 = \infty$ (where ∞ means infinity; Hindu mathematicians who introduced zero also did not consider this last proposition).

Gödel showed that for any system of mathematical logic, complex and consistent, there exist propositions expressed in the terminology of the system that can be neither proved nor disproved. When the system is enlarged and made more complex, however, and its internal consistency preserved, then previously undecidable propositions can be demonstrated to be either true or false. But for this system of enlarged complexity new and perhaps more numerous undecidable propositions now exist. It is always possible, given any system of mathematical logic, to formulate propositions that are decidable only within the context of larger and more complex systems.

Gödel's incompleteness theorem states that all quantifiable logical systems are incomplete. For example, statements of number theory occur in the *Principia Mathematica* whose truth cannot be established by its own system of logic.

Language, though less rigorous and complete, is more subtle and complex than mathematics and provides illustrations of Gödel's theorem. Thus in a town a male barber shaves all men who do not shave themselves. Who shaves the barber? Also, "this statement is false," exemplifies an undecidable proposition: if true, the statement is false; if false, it is true. Gödel pointed out the analogy between his theorem and the paradox of the liar. Epimenides said, "all Cretans are liars," yet was himself a Cretan. If he told the truth he lied because he was a Cretan; only if he lied could he have told the truth.

Many consistent and valid schemes of geometry have been developed and codified. Geometrical postulates by themselves, however, cannot determine whether the Euclidean or some other scheme is true or false when applied to the physical world. Not only in geometry but in every subject we find analogies to Gödel's theorem. When deciding on whether something is either true or false in one subject, we must often draw on supplementary knowledge in other subjects.

Gödel's incompleteness theorem reawakens old philosophical doubts. Can we justify reasoning using only the methods of reasoning?

It seems that we cannot. Also it creates new doubts. Can we justify the belief that the Universe itself is rational within the context of a particular rational universe? Can we identify ourselves as image makers within the imagery of our universe? And can we determine what is true and false in our universe without the advantage of referring to a more complex and complete universe? Again, it seems that we cannot.

We must complement our incomplete systems of deductive logic with inductive methods of inquiry, and endeavor to resolve all undecidable propositions by recourse to observation and experiment. We grope our way, hoping our guesses and hunches are right. Then along comes another universe in which most of our previous guesses and hunches turn out wrong.

□   □   □

Gödel's incompleteness theorem in deductive logic is analogous to Nicholas of Cusa's principle of learned ignorance. On the one hand, to every system of deductive logic there are undecidable propositions; on the other, to every system of knowledge there is an awareness of ignorance.

We think we know when unaware that we do not know. The possibility of knowledge has its roots in the soil of unlearned ignorance, which is another way of stating the third law. Ultimately, beyond all systems, stands the Universe in a cloud of unknowing. By love may it be gotten and holden; but by thought neither. He will be the more learned, the more he comes to know himself for ignorant.

BIBLIOGRAPHY

Many of the works cited are recent paperback editions. Further sources and readings may be found in the *Dictionary of Scientific Biography* (Scribner's, New York, 1972), the *Encyclopedia of Philosophy* (Macmillan, New York, 1967), and the eleventh edition of the *Encyclopedia Brittanica* (Cambridge University Press, 1910).

CHAPTER 1 Introducing the Masks

A. A. Arons and A. M. Bork, editors, *Science and Ideas: Selected Readings*. Prentice-Hall, Englewood Cliffs, New Jersey, 1964.

G. Berkeley, *A Treatise Concerning the Principles of Human Knowledge,* edited by C. M. Turbayne. The Library of Liberal Arts, Indianapolis, 1957.

C. Blakemore, *Mechanics of the Mind*. Cambridge University Press, New York, 1977.

F. H. C. Crick, "Thinking about the brain." *Scientific American,* September 1979.

A. S. Eddington, *Space, Time, and Gravitation*. Cambridge University Press, Cambridge, UK, 1920. Reprinted by Harper Torchbooks, New York, 1959.

S. Epstein, "The self-concept revisited." *American Psychologist,* **28,** 404, 1973.

A. Flew, editor, *Body, Mind, and Death*. Macmillan, New York, 1964.

A. Franknoi, "Images of the universe." *Mercury,* March–April 1975.

C. C. Gillispie, *The Edge of Objectivity*. Princeton University Press, Princeton, 1960.

R. L. Gregory, *Mind in Science: A History of Explanations in Psychology and Physics.* Cambridge University Press, New York, 1981.

B. Hoffmann, *Albert Einstein: Creator and Rebel*. Viking Press, New York, 1972.

D. H. Hubel, "The brain." *Scientific American,* September 1979.

T. H. Huxley, "Nature: Aphorisms by Goethe." *Nature,* **1,** 9, 1869.

M. Kline, *Mathematics in Western Culture*. Oxford University Press, New York, 1953.

W. MacKay, *Brains, Machines and Persons*. Eedmans, New York, 1980.

N. Malcolm, *Problems of the Mind: Descartes to Wittgenstein*. Harper and Row, New York, 1971.

M. K. Munitz, *Theories of the Universe: From Babylonian Myth to Modern Science.* Free Press, New York, 1957.

W. Penfield, *The Mystery of the Mind*. Princeton University Press, Princeton, 1975.

E. Schrödinger, *Mind and Matter*. Cambridge University Press, Cambridge, UK, 1958.

G. S. Stent, "Limits to the scientific understanding of man." *Science*, **187**, 1052, 1975.

CHAPTER 2 The Magic Universe

E. Durkheim, *Elementary Forms of the Religious Life*. Free Press, New York, 1965.

A. F. Elkin, *The Australian Aborigines*. The Natural History Library, Doubleday, Garden City, New York, 1964.

E. Evans-Pritchard, *Theories of Primitive Religion*. Oxford University Press, London, 1965.

E. Evans-Pritchard, *A History of Anthropological Thought*. Basic Books, New York, 1981.

P. Farb, *Humankind*. Houghton Mifflin, Boston, 1978.

H. Frankfurt, H. A. Frankfort, J. A. Wilson, and T. Jacobson, *Before Philosophy: The Intellectual Adventure of Ancient Man*. Penguin Books, Harmondsworth, Middlesex, UK, 1949.

J. G. Frazer, *The Golden Bough: A Study in Magic and Religion*, abridged edition. Macmillan, New York, 1945.

N. P. Hickerson, *Linguistic Anthropology*. Holt, Rinehart and Winston, New York, 1980.

W. Howells, editor, *Ideas on Human Evolution: Selected Essays*. Harvard University Press, Boston, 1962.

W. Howells, *Evolution of the Genus Homo*. Addison-Wesley, Reading, Massachusetts, 1973.

A. MacBeath, *Experiments in Living: A Study of the Nature and Foundations of Ethics or Morals in the Light of Recent Work in Social Anthropology*. Macmillan, London, 1952.

B. Malinowski, *Magic, Science and Religion and Other Essays*. Doubleday, Garden City, New York, 1948.

M. F. A. Montague, editor, *Culture and the Evolution of Man*. Oxford University Press, New York, 1962.

J. Napier, *The Roots of Mankind*. Harper and Row, New York, 1973.

J. Needham, editor, *Science, Religion and Reality*. George Braziller, New York, 1955.

P. H. Nowell-Smith, "Religion and Morality," in *The Encyclopedia of Philosophy*. Macmillan, New York, 1967.

P. Radin, *Primitive Religion, Its Nature and Origin*. Dover Publications, New York, 1957.

P. Radin, *Primitive Man as Philosopher*. Dover Publications, New York, 1957.

E. Sapir, *Language: An Introduction to the Study of Speech*. Harcourt, Brace and World, New York, 1921.

G. E. Swanson, *Birth of the Gods: The Origin of Primitive Beliefs*. University of Michigan Press, Ann Arbor, 1964.

W. D. Whitney, *The Life and Growth of Language*. Dover Publications, New York, 1979. First published in 1875.

CHAPTER 3 The Mythic Universe

R. Benedict, *Patterns of Culture*. New American Library, New York, 1948.

C. Blacker and N. Loewe, editors, *Ancient Cosmologies*. Allen and Unwin, London, 1975.

H. Butterfield, *The Origins of History*. Basic Books, New York, 1981.

J. Campbell, *The Hero with a Thousand Faces*. Princeton University Press, Princeton, 1949.

J. Campbell, *The Masks of God: Primitive Mythology*. Penguin Books, New York, 1976.

J. Campbell, *The Masks of God: Creative Mythology*. Penguin Books, New York, 1976.

V. G. Childe, *What Happened in History*. Penguin Books, Harmondsworth, Middlesex, UK, 1942.

M. Eliade, *A History of Religious Ideas: From the Stone Age to the Eleusian Mysteries*. University of Chicago Press, Chicago, 1978.

E. Hamilton, *Mythology: Timeless Tales of Gods and Heroes*. New American Library, 1969.

C. L. Hay, R. L. Linton, S. K. Lothrop, H. L. Shapiro, and G. C. Vaillant, editors, *The Maya and Their Neighbors*. Dover Publications, New York, 1977.

G. S. Kirk, *Myth, Its Meaning and Functions in Ancient and Other Cultures*. University of California Press, Berkeley, 1973.

M. K. Munitz, editor, *Theories of the Universe: From Babylonian Myth to Modern Science*. Free Press, New York, 1957.

N. K. Sanders, *The Epic of Gilgamesh*. Penguin Books, Harmondsworth, Middlesex, UK, 1972.

H. J. Schoeps, *The Religions of Mankind: Their Origin and Development*. Doubleday, New York, 1966.

B. C. Sproul, *Primal Myths: Creating the World*. Harper and Row, New York, 1979.

E. M. Thomas, *The Harmless People*. Random House, New York, 1958.

A. W. Watts, *The Two Hands of God*. Collier Books, New York, 1969.

A. W. Watts, *Myth and Ritual in Christianity*. Vanguard Press, New York, 1954.

A. N. Whitehead, *Science and the Modern World*. Macmillan, London, 1925.

CHAPTER 4 The Geometric Universe

R. E. Allen, *Greek Philosophy: Thales to Aristotle*. Free Press, New York, 1966.

J. Burnet, *Early Greek Philosophy*. Meridian Books, New York, 1957.

F. M. Cornford, *Plato's Cosmology*. Routledge and Kegan Paul, London, 1937.

F. M. Cornford, *From Religion to Philosophy*. Harper and Row, New York, 1957.

F. Cumont, *Astrology and Religion Among the Greeks and Romans*. Dover Publications, New York, 1960.

Diogenes Laertius, *Lives of Eminent Philosophers*, translated by R. D. Hicks, 2 volumes. Loeb Classical Library, New York, 1935.

A. Einstein, *The World as I See It*. The Philosophical Library, New York, 1934.

B. Farrington, *Greek Science, Its Meaning for Us*. Vol. 1, *Thales to Aristotle*; Vol. 2, *Theophrastus to Galen*. Penguin Books, Harmonsworth, Middlesex, UK, 1944.

G. Gale, *Theory of Science: An Introduction to the History, Logic, and Philosophy of Science*. McGraw-Hill, New York, 1979.

E. R. Harrison, *Cosmology: The Science of the Universe*. Cambridge University Press, New York, 1981. See Chapters 4, 5, and 6.

T. Heath, *Aristarchus of Samos: The Ancient Copernicus*. Clarendon Press, Oxford, UK, 1913.

R. Hope, *Aristotle's Physics*. University of Nebraska Press, Lincoln, 1961.

D. A. Hyland, *The Origins of Philosophy*. Putnam, New York, 1973.

C. S. Kirk and J. E. Raven, *The Presocratic Philosophers*. Cambridge University Press, Cambridge, UK, 1960.

T. S. Kuhn, *Copernican Revolution: Planetary Astronomy in the Development of Western Thought*. Harvard University Press, Cambridge, 1957.

Lucretius, *The Nature of the Universe*, translated by R. Latham. Penguin Books, Harmondworth, Middlesex, UK, 1951.

A. Mourelatos, *The Route of Parmenides*. Yale University Press, New Haven, 1970.

O. Neugebauer, *The Exact Sciences in Antiquity*. Brown University Press, Providence, 1957.

A. Pannekoek, *A History of Astronomy*. Interscience, New York, 1961.

Plato, *Timaeus*, in *The Collected Dialogues of Plato*, edited by E. Hamilton and H. Cairns. Pantheon, New York, 1961.

W. O. Roberts, "Science, a wellspring of our discontent." *American Scientist*, **55**, 3, 1967.

B. Russell, *A History of Western Philosophy*. Simon and Schuster, New York, 1972.

S. Sambursky, *The Physical World of the Greeks*. Routledge and Kegan Paul, London, 1956.

S. Sambursky, *Physics of the Stoics*. Routledge and Kegan Paul, London, 1959.

D. J. de Solla Price, *Science Since Babylon*. Yale University Press, New Haven, 1961.

G. Sarton, *Ancient Science: Through the Golden Age of Greece*. Oxford University Press, London, 1953.

J. L. Saunders, *Greek and Roman Philosophy after Aristotle*. Free Press, New York, 1966.

G. Vlastos, *Plato's Universe*. University of Washington Press, Seattle, 1975.

B. L. Van der Waerden, *Science Awakening*, translated by A. Dresden. Noordhoff, Groningen, Holland.

P. Wheelwright, *Aristotle: Containing Selections from Seven of the Most Important Books of Aristotle*. Odyssey Press, New York, 1951.

CHAPTER 5 The Medieval Universe

S. Angus, *The Mystery-Religions: A Study in the Religious Background of Early Christianity*. Dover Publications, New York, 1975.

W. C. Bark, *Origins of the Medieval World*. Stanford University Press, Stanford, 1958.

G. Barraclough, *Turning Points in World History*. Thames and Hudson, London, 1979.

C. L. Becker, *The Heavenly City of the Eighteenth-Century Philosophers*. Yale University Press, New Haven, 1968.

Boethius, *The Consolation of Philosophy*, translated by V. E. Watts. Penguin, Harmondsworth, Middlesex, UK, 1969.

J. B. Bury, *History of the Later Roman Empire*. Dover Publications, New York, 1958.

A. C. Crombie. *Augustine to Galileo*. Vol. 1, *Science in the Middle Ages 5th–13th Centuries;* Vol. 2, *Science in the Later Middle Ages and Early Modern Times 13th–17th Centuries*. Mercury Books, London, 1964.

F. Cumont, *The Mysteries of Mithra*. Dover Publications, New York, 1956.

Dante, *The Divine Comedy: Paradise,* translated by D. Sayers and B. Reynolds. Penquin Books, Harmondsworth, Middlesex, UK, 1962.

A. Freemantle, *The Age of Belief*. New American Library, New York, 1954.

J. A. Garraty and P. Gay, editors, *The Columbia History of the World*. Harper and Row, New York, 1983.

J. Gimpel, *The Medieval Machine: The Industrial Revolution of the Middle Ages*. Penguin Books, New York, 1977.

E. Grant, *Physical Science in the Middle Ages*. Cambridge University Press, New York, 1977.

A. Guillaume, *Islam*. Penguin Books, Harmondsworth, Middlesex, UK, 1954.

C. H. Haskins, *The Rise of Universities*. Cornell University Press, Ithaca, 1957.

S. K. Heninger, *The Cosmographical Glass: Renaissance Diagrams of the Universe*. Huntington Library, San Marino, California, 1977.

G. Holmes, *The Later Middle Ages 1272–1485*. Norton, New York, 1962.

A. Hyman and J. J. Walsh, editors, *Philosophy of the Middle Ages*. Hackett Publishing Company, Indianapolis, 1973.

K. Jaspers, *Anselm and Nicholas of Cusa*. Harcourt Brace Jovanovich, New York, 1974.

P. Johnson, *A History of Christianity*. Atheneum, New York, 1979.

G. Leff, *Medieval Thought*. Penguin Books, Harmondsworth, Middlesex, UK, 1956.

C. S. Lewis, *The Discarded Image: An Introduction to Medieval and Renaissance Literature*. Cambridge University Press, New York, 1964.

D. C. Lindberg, editor, *Science in the Middle Ages,* University of Chicago Press, Chicago, 1978.

A. O. Lovejoy, *The Great Chain of Being*. Harvard University Press, Cambridge, 1936.

S. H. Nasr, *An Introduction to Islamic Cosmological Doctrines*. Shambhala, Boulder, Colorado, 1978.

S. R. Packard, *12th Century Europe: An Interpretive Essay*. University of Massachusetts Press, Amherst, 1973.

O. Pedersen and M. Pihl, *Early Physics and Astronomy: A Historical Introduction*. American Elsevier, New York, 1974.

J. B. Ross and M. M. McLaughlin, *The Portable Medieval Reader*. Viking Press, New York, 1949.

G. Sarton, *Introduction to the History of Science*, Vol. 1, *From Homer to Omar Khayyam;* Vol. 2, *From Rabbi ben Ezra to Roger Bacon;* Vol. 3, *Science and Learning in the Fourteenth Century*. Williams and Wilkins, Baltimore, 1927.

W. W. Tarn, *Alexander the Great*. Beacon Press, Boston, 1956.

R. S. Westfall, *The Construction of Modern Science: Mechanism and Mechanics*. Cambridge University Press, New York, 1977.

L. White, *Medieval Technology and Social Change*. Oxford University Press, New York, 1964.

CHAPTER 6 The Infinite Universe

J. A. Bennett, "Christopher Wren: Astronomy, architecture and the mathematical sciences." *Journal for the History of Astronomy*, **6**, 149, 1975.

J. A. Bennett, "Cosmology and the magnetical philosophy, 1640–1680." *Journal for the History of Astronomy*, **12**, 165, 1981.

A. Berry, *A Short History of Astronomy: From Earliest Times Through the Nineteenth Century*. Dover Publications, New York, 1961.

I. B. Cohen, *The Newtonian Revolution*. Cambridge University Press, New York, 1981.

S. Drake, *Galileo at Work*. University of Chicago Press, Chicago, 1978.

A. R. Hall, *The Scientific Revolution 1500–1800*. Beacon Press, Boston, 1962.

A. R. Hall, *From Galileo to Newton*. Dover Publications, New York, 1981.

E. R. Harrison, "The dark night sky paradox." *American Journal of Physics*, **45**, 119, 1977.

E. R. Harrison, "The dark night sky riddle: a paradox that resisted solution." *Science*, **226**, 941, 1984.

N. S. Hetherington, "Man, society and the universe." *Mercury*, November–December, 1975.

M. A. Hoskin, "Newton, providence and the universe of stars." *Journal for the History of Astronomy*, **8**, 77, 1977.

C. Huygens, *The Celestial Worlds Discovered*. Frank Cass, London, 1968. First published in 1690.

S. L. Jaki, *The Paradox of Olbers' Paradox*. Herder, New York, 1969.

F. R. Johnson, "Gresham College: Precursor of the Royal Society." *Journal of the History of Ideas*, **1**, 413, 1940.

F. R. Johhson and S. V. Larkey, "Digges, the Copernican System, and the idea of the infinity of the universe in 1576." Huntington Library Bulletin, Harvard University Press, April 1934.

F. F. Jones, *Ancients and Moderns: A Study of the Rise of the Scientific Movement in Seventeenth-Century England*. Dover Publications, New York, 1982.

H. C. King, *The History of the Telescope*. Dover Publications, New York, 1979.

A. Koestler, *The Watershed: A Biography of Johannes Kepler*. Doubleday, Garden City, New York, 1960.

A. Koyré, *From the Closed World to the Infinite Universe*. Harper and Row, New York, 1958.

L. U. Pancheri, "Pierre Gassendi, a forgotten but important man in the history of physics." *American Journal of Physics*, **43**, 455, 1978.

W. Pauli, "The influence of archetypal ideas on the scientific theories of Kepler," in *The Interpretation of Nature and the Psyche*, translated by P. Silz. Pantheon, New York, 1955.

C. A. Ronan, *Edmond Halley: Genius in Eclipse*. Doubleday, Garden City, New York, 1969.

E. Rosen, *Kepler's Conversation with Galileo's Sidereal Messenger*. Johnson Reprint Corporation, New York, 1965.

G. de Santillana, *The Crime of Galileo*. University of Chicago Press, Chicago, 1955.

G. de Santillana, *The Age of Adventure*. New American Library, New York, 1956.

D. W. Singer, *Giordano Bruno, His Life and Thought*. Henry Schuman, New York, 1950.

E. G. R. Taylor, *The Mathematical Practitioners of Tudor and Stuart England*. Cambridge University Press, Cambridge, UK, 1954.

S. Toulmin and J. Goodfield, *The Fabric of the Heavens: The Development of Astronomy and Dynamics*. Harper and Row, New York, 1961.

N. Wade, "Thomas S. Kuhn: Revolutionary theorist of Science." *Science*, **197**, 144, 1977.

R. S. Westfall, *Never at Rest: A Biography of Isaac Newton*. Cambridge University Press, New York, 1981.

D. T. Whiteside, "Before the *Principia*: The maturing of Newton's thoughts on dynamical astronomy." *Journal for the History of Astronomy*, **1**, 5, 1970.

G. J. Whitrow, *The Structure and Evolution of the Universe: An Introduction to Cosomology*. Hutchinson, London, 1959.

CHAPTER 7 The Mechanistic Universe

C. L. Becker, *The Heavenly City of the Eighteenth-Century Philosophers*. Yale University Press, New Haven, 1968.

R. Berendzen, R. Hart, and D. Seeley, *Man Discovers the Galaxies*. Science History Publications, New York, 1976.

C. Brinton, *The Portable Age of Reason Reader*. Viking Press, New York, 1956.

J. D. Burchfield, *Lord Kelvin and the Age of the Earth*. Science History Publications, New York, 1975.

E. A. Burtt, *The Metaphysical Foundations of Modern Physical Science*. Doubleday, Garden City, New York, 1932.

C. Darwin, *The Origin of Species by Means of Natural Selection, or the Preservation of Favoured Races in the Struggle for Life*. Penguin Books, Harmondsworth, Middlesex, UK, 1968. First published by John Murray, 1859.

L. Eiseley, *The Immense Journey*. Vintage Books, New York, 1957.

A. Erman, *The Ancient Egyptians: A Sourcebook of their Writings*, translated by A. M. Blackman. Harper and Row, New York, 1966.

O. Gillie, *The Living Cell*. Funk and Wagnalls, New York, 1971.

S. J. Gould, *Ever Since Darwin: Reflections in Natural History*. Norton, New York, 1977.

J. C. Greene, *Darwin and the Modern World View*. Louisiana State University Press, Baton Rouge, 1961.

J. C. Greene, "Evolution and progress." *Johns Hopkins Magazine*, October 1962.

F. C. Haber, *The Age of the World: Moses to Darwin*. The Johns Hopkins Press, Baltimore, 1959.

E. Hansot, *Perfection and Progress: Two Modes of Utopian Thought*. M.I.T. Press, Cambridge, 1974.

M. A. Hoskin, *William Herschel and the Construction of the Heavens*. Science History Publications, New York, 1963.

M. A. Hoskin, "The cosmology of Thomas Wright of Durham." *Journal for the History of Astronomy*, **1**, 44, 1970.

P. James, *Population Malthus: His Life and Times*. Routledge and Kegan Paul, Boston, 1979.

I. Kant, *A Universal History and Theory of the Heavens*, translated by W. Hasties. University of Michigan Press, Ann Arbor, 1969.

E. Mayr, "The nature of the Darwinian revolution." *Science*, **176**, 981, 1972.

E. Mayr, "Darwin and natural selection." *American Scientist*, **65**, 321, 1977.

L. E. Orgel, *The Origins of Life: Molecules and Natural Selection*. Wiley, New York, 1973.

A. Pannekoek, *A History of Astronomy*. Interscience Publishers, New York, 1961.

R. S. Richardson, *The Star Lovers*. Macmillan, New York, 1967.

H. H. Ross, *Understanding Evolution*. Prentice-Hall, Englewood Cliffs, New Jersey, 1966.

S. Schaffer, "The phoenix of nature: fire and evolutionary cosmology in Wright and Kant." *Journal for the History of Astronomy*, **9**, 180, 1978.

E. Schrödinger, *What Is Life? The Physical Aspect of the Living Cell*. Cambridge University Press, New York, 1946.

J. B. Sidgwick, *William Herschel, Explorer of the Heavens*. Faber and Faber, London, 1955.

G. G. Simpson, *The Meaning of Evolution*. Yale University Press, New Haven, 1949.

G. G. Simpson, *This View of Life: The World of an Evolutionist*. Harcourt, Brace and World, New York, 1963.

L. Stephen, *History of English Thought in the Eighteenth Century*, 2 vols. Smith and Elder, London, 1881.

L. P. Williams, "Michael Faraday and the physics of 100 years ago." *Science*, **156**, 1335, 1967.

T. Wright, *An Original Theory or New Hypothesis of the Universe*, reprint of the 1750 edition, with an introduction by M. A. Hoskin. American Elsevier, New York, 1971.

CHAPTER 8 Dance of the Atoms and Waves

M. Berman, *The Reenchantment of the World*. Cornell University Press, Ithaca, 1982.

D. Bohm, *Quantum Theory*. Prentice-Hall, Englewood Cliffs, New Jersey, 1951.

V. Bush, *Science Is Not Enough*. Morrow, New York, 1967.

F. Capra, *The Turning Point: Science, Society and the Rising Culture*. Simon and Schuster, New York, 1982.

R. L. Carovillano and J. W. Skehan, editors, *Science and the Future of Man*. M.I.T. Press, Cambridge, 1970.

K. K. Darrow, "The quantum theory." *Scientific American*, May 1954.

P. C. W. Davies, *The Forces of Nature*. Cambridge University Press, Cambridge, UK, 1979.

P. A. M. Dirac, "The evolution of the physicist's picture of nature." *Scientific American*, May 1963.

S. D. Drell, "When is a particle?" *Physics Today*, June 1978.

F. Dyson, *Disturbing the Universe*. Harper and Row, New York, 1979.

A. S. Eddington, *The Internal Constitution of the Stars*. Dover Publications, New York, 1959. First published in 1926.

B. d'Espagnat, "The quantum theory and reality." *Scientific American*, November 1979.

G. Feinberg, *What Is the World Made of? Atoms, Leptons, Quarks, and Other Tantalizing Particles*, Doubleday, Garden City, New York, 1977.

R. Feynman, *The Character of Physical Law*. M.I.T. Press, Cambridge, 1967.

K. W. Ford, *The World of Elementary Particles*. Blaisdell, New York, 1965.

J. H. Gaisser, and T. K. Gaisser, "Partons in antiquity." *American Journal of Physics*, **45**, 439, 1977.

H. Georgi and S. L. Glashow, "Unified theory of elementary-particle forces." *Physics Today*, September 1980.

W. Heisenberg, *Physics and Philosophy: The Revolution in Modern Physics*. Harper and Row, New York, 1958.

B. Hoffmann, *The Strange Story of the Quantum*. Dover Publications, New York, 1959.

M. Jammer, *The Conceptual Development of Quantum Mechanics*. McGraw-Hill, New York, 1966.

G. Murchie, *Music of the Spheres: The Material Universe — From Atom to Quasar Simply Explained*; Vol. 1, *The Macrocosm: Planets, Stars, Galaxies, Cosmology*; Vol. 2, *The Microcosm: Matter, Atoms, Waves, Radiation, Relativity*. Dover Publications, New York, 1967.

H. R. Pagels, *The Cosmic Code: Quantum Physics as the Language of Nature*. Simon and Schuster, New York, 1982.

E. Schrödinger, *Science, Theory, and Man*. Dover Publications, New York, 1957.

E. Schrödinger, "What is matter?" *Scientific American*, September 1953.

A. G. Van Melsen, *From Atomos to Atom*. Harper, New York, 1960.

V. F. Weisskopf, *Knowledge and Wonder: The Natural World as Man Knows It*. Doubleday, Garden City, New York, 1963.

V. F. Weisskopf, "Three steps in the structure of matter." *Physics Today*, August 1970.

V. F. Weisskopf, "Of atoms, mountains and stars: A study in qualitative physics." *Science*, **187**, 605, 1975.

V. F. Weisskopf, "Is physics human?" *Physics Today*, June 1976.

CHAPTER 9 The Fabric of Space and Time

Augustine, *Confessions*, translated by R. S. Pine-Coffin. Penguin Books, Harmondsworth, Middlesex, UK, 1961.

E. Borel, *Space and Time*. Dover Publications, New York, 1960.

D. Bohm, *Causality and Chance in Modern Physics*, University of Pennsylvania Press, Philadelphia, 1971.

M. Born, *Natural Philosophy of Cause and Chance*. Clarendon Press, Oxford, UK, 1949.

M. Born, "Man and the atom." *Bulletin of the Atomic Scientists*, June 1957.

J. G. Crowther, "James Clerk Maxwell: Demon of modern physics." *New Scientist*, 8 March, 1979.

K. G. Denbigh, *An Inventive Universe*. George Braziller, New York, 1975.

J. W. Dunne, *An Experiment with Time*. Macmillan, New York, 1927.

A. S. Eddington, *The Nature of the Physical World*. Macmillan, New York, 1928.

A. Einstein and L. Infeld, *The Evolution of Physics from Early Concepts to Relativity and Quanta*. Simon and Schuster, New York, 1938.

M. Eliade, *The Myth of the Eternal Return*. Princeton University Press, Princeton, 1971.

J. T. Fraser, editor, *The Voices of Time*. University of Massachusetts Press, Amherst, 1981.

R. M. Gale, editor, *The Philosophy of Time: A Collection of Essays*. Humanities Press, Atlantic Highlands, New Jersey, 1968.

M. Gardner, *The Ambidextrous Universe*. Basic Books, New York, 1964.

T. Gold, editor, *The Nature of Time*. Cornell University Press, Ithaca, 1967.

E. T. Hall, *The Hidden Dimension*. Doubleday, Garden City, New York, 1969.

E. R. Harrison, *Cosmology: The Science of the Universe*. Cambridge University Press, New York, 1981. See Chapter 6.

I. Hinckfuss, *The Existence of Space and Time*. Clarendon Press, Oxford, 1975.

C. H. Hinton, *Speculations on the Fourth Dimension: Selected Writings*, edited by R. B. Rucker. Dover Publications, New York, 1980.

S. Hook, editor, *Determinism and Freedom in the Age of Modern Science*. Macmillan, New York, 1961.

M. Jammer, *Concepts of Space*. Harper and Brothers, New York, 1960.

A. Koslow, *The Changeless Order: The Physics of Space, Time and Motion*. George Braziller, New York, 1967.

C. Lanczos, *Space Through the Ages*. Academic Press, New York, 1970.

J. Needham, *Time: The Refreshing River*. Allen and Unwin, London, 1943.

J. B. Priestley, *Man and Time*. Dell, New York, 1968.

H. Reichenbach, *The Philosophy of Space and Time*. Dover Publications, New York, 1957.

H. Reichenbach, *The Direction of Time*. University of California Press, Berkeley, 1971.

J. J. C. Smart, editor, *Problems of Space and Time: From Augustine to Albert Einstein*. Macmillan, New York, 1964.

S. Toulmin and J. Goodfield, *The Discovery of Time*. Harper and Row, New York, 1965.

H. G. Wells, *The Time Machine*. Random House, New York, 1931. Reprinted with an introduction by Wells.

H. Weyl, *Philosophy of Mathematics and Natural Science*. Princeton University Press, Princeton, 1949.

J. A. Wheeler, "The computer and the universe." *International Journal of Theoretical Physics*, **21**, 557, 1982.

G. J. Whitrow, *What Is Time?* Thames and Hudson, London, 1972.

G. J. Whitrow, *The Natural Philosophy of Time*, second edition. Clarendon Press, Oxford, UK, 1980.

E. Whittaker, *From Euclid to Eddington: A Study of Conceptions of the External World*. Dover Publications, New York, 1958.

CHAPTER 10 Nearer to the Heart's Desire

E. A. Abbott, *Flatland*. Dover Publications, New York, 1952.

J. Bernstein, *Einstein*. Viking Press, New York, 1973.

D. Burger, *Sphereland: A Fantasy about Curved Spaces and an Expanding Universe*. Crowell, New York, 1969.

W. Clifford, *The Common Sense of the Exact Sciences*. Dover Publications, New York, 1955.

P. C. W. Davies, *Space and Time in the Modern Universe*. Cambridge University Press, Cambridge, UK, 1977.

A. S. Eddington, *The Nature of the Physical World*. Cambridge University Press, Cambridge, UK, 1928.

A. S. Eddington, *Space, Time, and Gravitation*. Harper and Row, New York, 1959.

A. Einstein, *Out of My Later Years*. Citadel Press, Secaucus, New Jersey, 1974.

A. Einstein, *The World as I See It*, translated by A. Harris. Citadel Press, Secaucus, New Jersey, 1979.

G. F. R. Ellis, "Cosmology and verifiability." *Quarterly Journal of the Royal Astronomical Society*, **16**, 245, 1975.

P. Frank, *Einstein: His Life and Times*. Knopf, New York, 1947.

E. R. Harrison, *Cosmology: The Science of the Universe*. Cambridge University Press, New York, 1981. See Chapters 7, 8, and 9.

B. Hoffman, *Albert Einstein: Creator and Rebel*. Viking Press, New York, 1972.

L. Infeld, *Albert Einstein: His Work and Influence on Our World Order*. Scribners, New York, 1950.

M. P. Jaggi, "The visionary ideas of Bernhard Riemann." *Physics Today*. December 1967.

C. Lanczos, *Albert Einstein and the Cosmic World Order*. Wiley, New York, 1965.

A. J. Meadows, *Stellar Evolution*. Pergamon Press, London, 1967.

A. Pais, *Subtle Is the Lord: The Science and the Life of Albert Einstein*. Oxford University Press, New York, 1983.

R. Penrose, "Black holes: new horizons in gravitational theory." *American Scientist*, September–October 1974.

S. Schaffer, "John Mitchell and black holes." *Journal for the History of Astronomy*, **10,** 42, 1979.

P. A. Schilpp, editor, *Albert Einstein: Philosopher-Scientist*. Library of Living Philosophers, Evanston, Illinois, 1949.

L. L. Smarr and W. H. Press, "Our elastic spacetime: Black holes and gravitational waves." *American Scientist*, January–February 1978.

I. Stewart, "Gauss," *Scientific American*, July 1977.

W. Sullivan, *Black Holes*. Doubleday, Garden City, New York, 1979.

J. A. Wheeler, "Our universe: The known and the unknown." *American Scientist*, Spring 1968.

J. C. Wheeler, "After the supernova, what?" *American Scientist*, January–February 1973.

G. J. Whitrow, editor, *Einstein: The Man and His Achievement*. Dover Publications, New York, 1973.

CHAPTER 11 The Cosmic Tide

H. Bondi, *Cosmology*. Cambridge University Press, Cambridge, UK, 1960.

P. C. W. Davies, *The Runaway Universe*. Dent, London, 1978.

A. S. Eddington, *The Expanding Universe*. Cambridge University Press, Cambridge, UK, 1933. Reprinted by the University of Michigan Press, Ann Arbor, 1958.

G. F. R. Ellis and G. B. Brundrit, "Life in the Infinite Universe." *Quarterly Journal of the Royal Astronomical Society*, **20,** 37, 1979.

G. Gamow, *The Creation of the Universe*. Viking, New York, 1952.

E. R. Harrison, *Cosmology: The Science of the Universe*. Cambridge University Press, New York, 1981. See Chapters 4, 10, and 11.

F. Hoyle, *The Nature of the Universe*. Penguin Books, Harmondsworth, Middlesex, UK, 1960.

I. R. King, *The Universe Unfolding*. Freeman, San Francisco, 1976.

W. H. McCrea, "Willem de Sitter, 1872–1934." *Journal of the British Astronomical Association*, **82**, 178, 1972.

J. D. North, *The Measure of the Universe: A History of Modern Cosmology*. Clarendon Press, Oxford, UK, 1964.

E. A. Poe, *Eureka*, reproduced in *The Science Fiction of Edgar Allen Poe*, edited by H. Beaver. Penguin Books, Harmondsworth, Middlesex, UK, 1976.

D. W. Sciama, *Modern Cosmology*. Cambridge University Press, Cambridge, UK, 1971.

G. T. Whitrow, *The Structure and Evolution of the Universe*. Hutchinson, London, 1959.

CHAPTER 12 Do Dreams Ever Come True?

G. T. Bath, editor, *The State of the Universe*. Clarendon Press, Oxford, 1980.

G. Burbidge, "Was there really a big bang?" *Nature*, **233**, 36, 1971.

F. Close, "The search for proton decay." *Nature*, **292**, 799, 1981.

F. Crick, *Life Itself: Its Origin and Nature*. Simon and Schuster, New York, 1982.

F. J. Dyson, "The future of physics." *Physics Today*, September 1970.

A. S. Eddington, "The beginning of the world from the point of view of quantum theory." *Nature*, March 31, 1931.

A. S. Eddington, "The end of the world from the standpoint of mathematical physics." *Nature*, Supplement, March 24, 1931.

G. F. R. Ellis and G. B. Brundrit, "Life in the infinite universe." *Quarterly Journal of the Royal Astronomical Society*, **20**, 37, 1974.

G. Gamow, *The Creation of the Universe*. Viking Press, New York, 1952.

E. R. Harrison, *Cosmology: The Science of the Universe*. Cambridge University Press, New York, 1981. See Chapters 11 and 18.

S. Hawking, "The quantum mechanics of black holes." *Scientific American*, January 1977.

K. Heur, *The End of the World: A Scientific Inquiry*. Victor Gollancz, London, 1953.

F. Hoyle, *The Nature of the Universe*. Penguin Books, Harmondsworth, Middlesex, UK, 1960.

J. N. Islam, "Possible ultimate fate of the universe." *Quarterly Journal of the Royal Astronomical Society*, **18**, 3, 1977. "The ultimate fate of the universe." *Sky and Telescope*, January 1979.

E. Kasner and J. Newman, *Mathematics and the Imagination*. Simon and Schuster, New York, 1952.

G. Lemaître, *The Primeval Atom*. Van Nostrand, New York, 1951.

Y. Nambu, "The confinement of quarks." *Scientific American*, November 1976.

J. Oro, S. L. Miller, C. Ponnamperuma, and R. S. Young, *Cosmochemical Evolution and the Origins of Life*. Reidel, Dordrecht, Holland, 1974.

E. A. Poe, *Eureka*, reproduced in *The Science Fiction of Edgar Allen Poe*, edited by H. Beaver. Penguin Books, Harmondsworth, Middlesex, UK, 1976.

C. Ponnamperuma, *The Origins of Life*. Dutton, New York, 1972.

M. Rees, "The collapse of the universe: An eschatological study." *Observatory*, **89,** 193, 1969.

A. L. Robinson, "Is a diamond really forever?" *Science,* **206,** 670, 1979.

A. Schuster, "Potential matter — a holiday dream." *Nature,* **58,** 367, 618, 1898.

D. W. Sciama, "The ether transmogrified." *New Scientist,* February 2, 1978.

J. Silk, *The Big Bang*. Freeman, San Francisco, 1979.

M. S. Turner and D. N. Schramm, "Cosmology and elementary-particle physics." *Physics Today*, September 1979.

R. Wald, Particle creation near black holes." *American Scientist*, September–October 1977.

M. M. Waldrop, "Inflation and the mysteries of the cosmos." *Science,* **213,** 121, 1981.

S. Weinberg, "Unified theories of elementary particle interactions." *Scientific American*, July 1974.

S. Weinberg, *The First Three Minutes: A Modern View of the Origin of the Universe*. Basic Books, New York, 1977.

S. Weinberg, "The decay of the proton." *Scientific American*, June 1981.

J. A. Wheeler, "The universe as a home for man." *American Scientist*, November–December 1974.

E. T. Whittaker, *The Beginning and End of the World*. Oxford University Press, London, 1942.

CHAPTER 13 The Witch Universe

H. Butterfield, *The Origins of Modern Science*. Free Press, New York, 1965.

R. Cavendish, *Visions of Heaven and Hell*. Harmony Books, New York, 1977.

N. Cohn, *Europe's Inner Demons: An Enquiry Inspired by the Great Witch Hunt*. Meridian Book, New American Library, New York, 1977.

R. S. Dunn, *The Age of Religious Wars, 1559–1715*. Norton, New York, 1979.

A. S. Eddington, *The Expanding Universe*. Cambridge University Press, Cambridge, UK, 1933. Reprinted by University of Michigan Press, Ann Arbor, 1958.

B. Halstad, "Popper: Good philosophy, bad science?" *New Scientist*, 17 July 1980.

H. Kramer and J. Sprenger, *The Malleus Maleficarum*, translated by M. Summers. Dover publications, New York, 1971. In the Introduction of the 1928 edition, Reverend Summers writes of the Holy Inquisition, "There can be no doubt that had this most excellent tribunal continued to enjoy its full perogative and the full exercise of its salutary powers, the world at large would be in a far happier and a far more orderly position today."

B. Magee, "Karl Popper: The useful philosopher." *Horizon*, Autumn 1974.

M. Marwick, editor, *Witchcraft and Sorcery*. Penguin Books, Harmondsworth, Middlesex, UK, 1970.

K. R. Popper, *The Logic of Scientific Discovery*. Harper and Row, New York, 1965.

K R. Popper, "Conjectures and refutations," in *Science: Men, Methods, Goals*, edited by B. A. Brody and N. Capaldi. Benjamim, New York, 1968.

H. R. Trevor-Roper, *The European Witch-Craze of the Sixteenth and Seventeenth Centuries and Other Essays*. Harper and Row, New York, 1969.

J. Ziman, *Reliable Knowledge: An Exploration of the Grounds for Belief in Science*. Cambridge University Press, Cambridge, UK, 1978.

CHAPTER 14 The Spear of Archytas

E. Anscombe and P. T. Geach, editors, *Descartes: Philosophical Writings*. Nelson, Edinburgh, 1954.

M. Aurelius, *Meditations*, translated by M. Staniforth. Penguin Books, Harmondsworth, Middlesex, UK, 1964.

H. Bondi, *Cosmology*. Cambridge University Press, Cambridge, UK, 1960.

H. Bondi, W. B. Bonnor, R. A. Lyttleton, and G. J. Whitrow, *Rival Theories of Cosmology*. Oxford University Press, London, 1960.

M. Gardner, *Fads and Fallacies in the Name of Science*. Dover Publications, New York, 1957.

F. C. Haber, *The Age of the World: Moses to Darwin*. Johns Hopkins Press, Baltimore, 1959.

E. R. Harrison, "The dark night sky paradox." *American Journal of Physics*, **45,** 119, 1977.

E. R. Harrison, *Cosmology: The Science of the Universe*. Cambridge University Press, New York, 1981. See Chapters 4, 5, and 15.

E. R. Harrison, "The dark night sky riddle: A paradox that resisted solution." *Science*, **226,** 941, 1984.

S. L. Jaki, *The Paradox of Olbers' Paradox*. Herder, New York, 1969.

A. Koyré, *Newtonian Studies*. Chapman and Hall, London, 1965.

J. Laurie, editor, *Cosmology Now*. British Broadcasting Corporation, London, 1974.

Lucretius, *The Nature of the Universe*, translated by R. Latham. Penguin Books, Harmondsworth, Middlesex, UK, 1951.

H. Margenau, *The Nature of Physical Reality: A Philosophy of Modern Physics*. McGraw-Hill, New York, 1950.

M. K. Munitz, *Space, Time and Creation*. Free Press, Glencoe, Illinois, 1957.

L. Rosenfeld, "Niels Bohr's Contribution to Epistemology." *Physics Today*, October 1963. Refers to Poul Møller's Danish student.

E. Schrödinger, *Mind and Matter*. Cambridge University Press, Cambridge, UK, 1958.

CHAPTER 15 All That Is Made

G. T. Bath, editor, *The State of the Universe*. Clarendon Press, Oxford, 1980.

J. D. Barrow and F. J. Tipler, *The Anthropic Principle*. Oxford University Press, London, 1982.

D. Bohm, *Wholeness and the Implicate Order*. Routledge and Kegan Paul, New York, 1980.

J. L. Borges, "The Garden of Forking Paths," in *Labyrinths: Selected Stories and Other Writings*, edited by D. A. Yates and J. E Irby. New Directions, Norfolk, Connecticut, 1962.

J. Buchler, *The Philosophical Writings of Peirce*. Dover Publications, New York, 1955.

B. J. Carr and M. J. Rees, "The anthropic principle and the structure of the physical world." *Nature*, **278,** 605, 1979.

B. Carter, "Large number coincidences and the anthropic principle in cosmology," in *Confrontations of Cosmological Theories with Observational Data*, edited by M. S. Longair. Reidel, Dordrecht-Holland, 1974.

G. F. Chew, " 'Bootstrap:' A scientific idea?" *Science*, **161**, 762, 1968.

R. G. Colodny, *Paradigms and Paradoxes: The Philosophical Challenge of the Quantum Domain*. University of Pittsburgh Press, Pittsburgh, 1972.

P. C. W. Davies, *Other Worlds*. Simon and Schuster, New York, 1980.

P. C. W. Davies, "The anthropic principle and the early universe." *Mercury*, May–June 1981.

P. C. W. Davies, *The Accidental Universe*. Cambridge University Press, Cambridge, UK, 1982.

B. S. DeWitt, "Quantum mechanics and reality." *Physics Today*, September 1970.

B. S. DeWitt and N. Graham, *The Many-Worlds Interpretation of Quantum Mechanics*. Princeton University Press, Princeton, 1973.

R. H. Dicke, "Dirac's cosmology and Mach's principle." *Nature*, **192**, 440, 1961.

P. A. M. Dirac, "A new basis for cosmology." *Nature*, **139**, 323, 1937.

G. F. R. Ellis, "The world's environment: The universe." *South African Journal of Science*, **75**, 529, 1979.

E. R. Harrison, "The cosmic numbers." *Physics Today*, December 1972.

E. R. Harrison, *Cosmology: The Science of the Universe*. Cambridge University Press, New York, 1981. See Chapters 5 and 17.

S. W. Hawking, *Is the End in Sight for Theoretical Physics?* Cambridge University Press, Cambridge, 1980.

L. J. Henderson, *The Fitness of the Environment*. Macmillan, New York, 1913.

Julian of Norwich, *Revelations of Divine Love*, translated into modern English by C. Wolters. Penguin Books, Baltimore, 1966.

J. Kidd, *The Bridgewater Treatises, on the Power, Wisdom, and Goodness of God, as Manifested in the Creation, on the Adaptation of External Nature to the Physical Condition of Man*. H. G. Bohn, London, sixth edition, 1852.

M. Maimonides, *The Guide for the Perplexed*, translated by M. Friedlander, Dover Publications, New York, 1956.

F. Mason and J. A. Thomson, *The Great Design: Order and Progress in Nature*. Duckworth, London, 1934.

G. R. Montgomery, *Leibniz: Basic Writings*. Open Court, La Salle, Illinois, 1962.

W. Paley, *Natural Theology*, edited by F. Ferre. Bobbs-Merrill, New York, 1963.

C. M. Patton and J. A. Wheeler, "Is Physics Legislated by Cosmology?" in *Quantum Gravity: An Oxford Symposium*, edited by C. J. Isham, R. Penrose, and D. W. Sciama. Clarendon Press, Oxford, UK, 1975.

M. J. Rees, "Our universe and others." *Quarterly Journal of the Royal Astronomical Society*, **22**, 109, 1981.

G. Sarton, *Introduction to the History of Science*, Vol. 1, *From Homer to Omar Khayyam*. Williams and Wilkins, Baltimore, 1927.

R. Schlegel, *Inquiry into Science: Its Domain and Limits*. Doubleday, Garden City, New York, 1972.

O. Stapledon, *Last and First Men and Star Maker*. Dover Publications, New York, 1968.

J. W. N. Sullivan, *The Limitations of Science*. Viking, New York, 1933.

V. Trimble, "Cosmology: Man's place in the universe." *American Scientist*, January–February 1977.

A. N. Whitehead, *The Adventure of Ideas*. Free Press, New York, 1933.

J. A. Wheeler, "The universe as a home for man." *American Scientist*, **62**, 683, 1974.

J. A. Wheeler, "Genesis and observership," in *Foundational Problems in the Special Sciences*, edited by R. E. Butts and J. Hintikka. Reidel, Dordrecht, Holland, 1977.

J. A. Wheeler, "Some Strangeness in the Proportion," in *A Centennial Symposium to Celebrate the Achievements of Albert Einstein*, edited by H. Woolf. Addison-Wesley, Reading, Massachusetts, 1980.

CHAPTER 16 The Cloud of Unknowing

M. J. Adler, *How to Think about God*. Macmillan, New York, 1980.

Anon., *Cloud of Unknowing, and Other Treatises*, edited by J. McCann, Newman Press, Westminster, Maryland, 1952.

V. J. Bourke, *The Pocket Aquinas*. Simon and Schuster, New York, 1960.

R. Descartes, *Discourse on Method and the Meditations*, translated by F. E. Sutcliffe. Penguin Books, Harmondsworth, Middlesex, UK, 1968.

A. Einstein, *The World as I See It*. The Philosophical Library, New York, 1934.

A. Flew, editor, *Body, Mind and Death*. Macmillan, New York, 1964.

E. Fromm and R. Xirau, editors, *The Nature of Man*. Macmillan, New York, 1968.

J. Goethe, quoted by T. Huxley, "Nature: Aphorisms by Goethe." *Nature*, **1**, 9, 1869.

J. Hick, editor, *The Existence of God*. Macmillan, New York, 1964. Contains a selection of original essays and an informative introduction.

K. Jaspers, *Anselm and Nicholas of Cusa*, edited by H. Arendt. Harcourt Brace Jovanovich, New York, 1966.

J. Laird, *Theism and Cosmology*. The Philosophical Library, 1942, reprinted by Books for Libraries Press, Freeport, New York, 1969.

B. Russell, *Why I Am Not a Christian*. Simon and Schuster, New York, 1957.

S. Weinberg, *The First Three Minutes*. Basic Books, New York, 1977.

A. N. Whitehead, *Science and the Modern World*. Cambridge University Press, Cambridge, UK, 1926.

E. Whittaker, *Space and Spirit: Theories of the Universe and the Arguments for the Existence of God*. Thomas Nelson and Sons, London, 1946.

CHAPTER 17 Learned Ignorance

H. Delong, *A Profile of Mathematical Knowledge*. Addison-Wesley, Reading, Massachusetts, 1970.

R. Dubos, *The Dreams of Reason*. Columbia University Press, New York, 1961.

P. Farb, "How do I know you meant what you mean?" *Horizon*, Autumn 1968.

K. Gödel, *On Formally Undecidable Propositions*. Basic Books, New York, 1962.

D. R. Hofstadter, *Gödel, Escher, and Bach: An Eternal Golden Braid*. Vintage Books, New York, 1979.

K. Jaspers, *Anselm and Nicholas of Cusa*, edited by H. Arendt. Harcourt Brace Jovanovich, New York, 1966.

A. Koyré, *From the Closed World to the Infinite Universe*. Harper and Row, New York, 1958.

A. Mostowski, *Sentences Undecidable in Formalized Arithmetic: An Exposition of the Theory of Kurt Gödel*. North-Holland, Amsterdam, 1957.

E. Nagel and J. R. Newman, *Gödel's Proof*. New York University Press, New York, 1958.

B. Russell, *Human Knowledge, Its Scope and Limits*. Allen and Unwin, London, 1948.

B. Russell, *The Problems of Philosophy*. Oxford University Press, London, 1959.

B. Stewart and P. G. Tait, *The Unseen Universe*. Macmillan, London, 1886.

L. Thomas, "On the uncertainty of science." *The Key Reporter*, Autumn 1980.

H. Weyl, *Philosophy of Mathematics and Natural Science*. Princeton University Press, Princeton, 1949.

Index

crystal ball, 153
Curveland, 161
Cuvier, Georges (1769–1832), 111

Dalton, John (1766–1844), 122
Dante (1265–1321), 37, 76, 77–79, 110
Dark Ages, 71, 214
Darwin, Charles (1809–1882), 114
Darwinian Revolution, 173
decay of matter, 196
deceleration, 181
deism, 102, 241
deistic principle, 256
Democritus (b. 460? B.C.), 55, 65, 222
Descartes, René (1596–1650), 92, 96, 101, 122, 141, 166, 235, 260–261
design argument, 248
determinism, 154, 236
deuterium, 192, 195
deuterons, 130
DeWitt, Bryce, 246
Dicke, Robert, 248–249
Dickens, Charles (1812–1870), 231
Dickinson, Emily (1830–1886), 166, 201
Diderot, Denis (1713–1784), 111
Digges, Thomas (1546?–1595), 89, 224
dinosaurs, 188
Diogenes Laertius (3rd century A.D.), 56
Dionysus, 57
Diophantus of Alexandria (3rd century A.D.), 65
Dirac, Paul (1902–1984), 194, 248
Dobson, Austin (1840–1921), 137
Dominicans, 72, 208
Doomsday, 39, 105
dreamship, 191
dreamtime, 137
Dunne, John, 138
Durkheim, Émile (1858–1917), 31

early universe, 176, 192
eccentric orbits, 64
Eddington, Sir Arthur (1882–1944), 13–14, 128–129, 152, 158, 178, 184, 198, 217, 229, 248
Einstein, Albert (1879–1955), 57, 142, 144, 148, 157, 160, 162, 163, 171, 174, 177, 196, 227, 271

Einstein equation, 164
Elda Edda, 39
Eleatics, 53
electricity, 94
electromagnetic theory, 141, 142
electron waves, 123–125
electroweak force, 196
electrons, 122, 124, 193
elliptical orbits, 97, 166
Empedocles (c. 490–430 B.C.), 54, 222
empiricism, 216
empyrean, 73
energy and mass, 165
Enkidu, 39–40
Enuma Elish, 38
Epicureans, 239
epicyclic orbits, 64
Epimenides, 277
Epimetheus, 39
Epstein, Seymour, 9
equivalence principle, 158–159
Erasmus, Desiderius (c. 1466–1536), 207
Eratosthenes of Alexander (276?–194? B.C.), 60
Eros, 39
Eternal City, 106
eternal return, 137
eternity, 202
ether, 61, 141
Euclid (c. 300 B.C.), 56, 64, 160
Euclidean geometry, 160
event horizon, 168
events, defined, 140
Everett, Hugh, 244
evolution, 112, 116
expanding universe, 176, 178
extreme early universe, 193

falsifiable knowledge, 216–219
falsified knowledge, 219
fate, 153
Feynmann, Richard, 124
Fielding, Henry (1707–1754), 250
fission, 131–132
FitzGerald, George (1851–1901), 142
Flatland, 161
Fontenelle, Bernard de (1657–1757), 101
fortune teller, 153
forces, angelic, 85
  astral, 92
  electrical, 121

night sky, 74
now, 138, 155, 156
Nowell-Smith, P.H., 30
nuclear energy, 128–133
nuclear reactions, 131
nuclear reactor, 131, 192
nuclear weapons, 131
nucleotides, 115, 190
nucleus of atom, 122, 127

observable universe, 184, 186, 197
Ockham, William of (c. 1280–1349), 84
Ockham's razor, 84, 267
Odin, 39
Old Testament, 43
ontological argument, 259–261
Oresme, Nicole (c. 1320–1382), 85–86, 93
origin of, galaxies, 190–191
  homogeneity, 198
  life, 189–190
  Solar System, 189
original sin, 69, 153, 237
Osiris, 44
ozone layer, 188

Paine, Thomas (1737–1809), 104
paleocosmic eras, 191
Paleozoic era, 188
Paley, William (1743–1805), 241
Pandora, 39
pantheism, 267–269
Paracelsus (1490–1541), 80
parallel postulate, 160
Parmenidean stillness, 153
Parmenides (6th–5th century B.C.), 53
Pascal, Blaise (1623–1662), 101
Pauli, Wolfgang (1900–1958), 91
peculiar velocity, 180, 183
Pelagian heresy, 154
Pelagius (late 4th to early 5th century), 154, 237
Penzias, Arno, 176
Pericles (c. 495–429 B.C.), 55
persecution of Jews, 207
Philoponus, John (5th–6th century A.D.), 85
philosopher's stone, 206
photons, 125, 127, 128, 134, 193, 201
Planck, Max (1858–1947), 126

Planck epoch, 198
  period, 197
Plato (428–348 B.C.), 52, 59, 61, 63, 222
Platonic system, 63
Pleistocene, 187
plenitude, 81
plurality of worlds, 86
plutonium, 131
Podsnap flourish, 231
Poe, Edgar Allan (1809–1849), 176, 200
polytheism, 267
Pope, Alexander (1688–1744), 81
popes, Alexander IV, 208
  Innocent VIII, 209
Popper, Karl, 216
population growth, 115, 132
positrons, 193, 194
predestination, 237
primordial Earth, 189
primordial Sun, 189
primum mobile, 73
Principia, 97, 137
principle of, containment, 221–222
  equivalence, 158–159
  uncertainty, 125–127
probability, 124, 245
progress, 116
Prometheus, 39
Protagoras (c. 490–after 421), 58
proteins, 190
proterocosmos, 193
Proterozoic era, 188, 190
Protestantism, 45
protons, 127, 192
Prout, William (1785–1850), 243
Pseudo-Dionysius, 76
psychology, 104
Ptolemaic system, 65, 87
Ptolemy, Claudius (2nd century A.D.), 64, 65
pulsars, 166, 169
purgatory, 74
Pythagoras (6th century B.C.), 56–58, 60
Pythagorean theorem, 148
Pythagoreans, 57, 63

Quakerism, 45
quantum black holes, 197
quantum mechanics, 122, 124

universe (*continued*)
  two-sphere, 60
  violent, 187
  whirligig, 137
  witch, 210
  Wright's, 173
  Zoroastrian, 232
universes, 266–271, 275
  defined, 1–2
  ensemble of, 249–253
  falsifiable, 217
  mad, 10
  magicomythic, 26–30
universities, 72
unlearned ignorance, 273 et seq.
Uranus, 39
Ussher, James (1581–1656), 110
Utnapishtim, 40

vacuum, 92, 141
Vandals, 68
Vatican Council, 258
velocity, 85
velocity-distance law, 179–186
Vesalius, Andreas (1514–1564), 80, 90
Vinci, Leonardo da (1452–1519), 85
Virgil (70–19 B.C.), 131
vitalists, 234
void, 92, 98, 222
Voltaire (1694–1778), 17, 74, 102

Waldenses, 208
Wallace, Alfred (1823–1913), 114
water, 243
Watts, Alan, 40, 44
wavelength, 184–185
weak interaction, 130
weight, 93
weightless, 158

Weinberg, Steven, 196, 271
Wells, H. G. (1866–1946), 142
Wheel of Time, 105, 137, 138, 155
Wheeler, John Archibald, 246
White, Lynn, 71
white dwarfs, 131, 168
White Queen, 199
Whitehead, Alfred (1861–1947), 45, 253, 271, 276
why-type question, 247, 255
Wilson, John, 20
Wilson, Robert, 176
Wisdom Literature, 43–44
witchcraft, 207
witches, 104, 209–215
witch craze, 209–214
Wordsworth, William (1770–1850), 106
world lines, defined, 140–141
world pictures, 48, 109, 110, 237
  defined, 8–9
  mad, 9–10
Wren, Sir Christopher (1632–1723), 95–96
Wright, Thomas (1711–1786), 107
Wurtz, Carl, 178

Xenophanus (6th–5th century B.C.), 222
X-rays, 170
X-ray sources, 170

Yahweh, 43
Ymir, 39

Zeus, 39
Zarathrusta, 43
Zoroaster (7th–6th century B.C.), 43, 50
Zoroastrianism, 44